MW01641412

ENGINEERING FOR THE OFFICER OF THE DECK

DANIEL G. FELGER
COMMANDER, U. S. NAVY

ENGINEERING FOR THE OFFICER OF THE DECK

NAVAL INSTITUTE PRESS
ANNAPOLIS, MARYLAND

Library of Congress Catalogue Card No. 78-70964
ISBN 0-87021-172-2

Printed in the United States of America

Contents

Foreword

The primary objective of this book is to assist you in becoming more familiar—and perhaps even "comfortable"—with the naval propulsion plants that provide the motive power for the great majority of our surface warships.

As a junior officer, you want to qualify as an officer of the deck (OOD) on board your ship as quickly as possible. A basic understanding of the propulsion plant of your ship is an important requisite in your qualification. If you haven't reported aboard ship yet, the whole idea of the engineering plant—what is "down there" and what it does—may be a mystery. Regardless of whether you are on board your ship now or still in the classroom, this book will aid you in achieving a basic understanding of the typical marine propulsion plants that power our surface warships.

Of course, qualification as a surface warfare officer (SWO) and an OOD comes only with the Captain's approval. Before you qualify you will spend many hours crawling over and around generators, turbines, and pumps, discussing their operation with the men who operate and maintain them, and reading numerous operating and technical manuals.

This book will neither make you an expert nor tell you everything you need to know about the engineering plant in your ship. It serves as no substitute for those numerous and necessary publications, the "skull" sessions with the men often referred to as "snipes," and the time you will spend on the deckplates in the engineering plant.

However, this book will help you to understand key *relationships* concerning the engineering plant of your ship. The majority of engineering publications available to junior naval personnel either presume a previous familiarity with the subject or else describe the function and operation of key equipments without devoting adequate explanation to the relationships between components. The engineer officer of your ship obviously needs to know a great deal about systems and components within the systems. As an OOD or a prospective OOD, however, you must be more concerned with systems and relationships than with the operation of a specific piece of equipment, such as the main feed pump in a conventional steam propulsion plant.

The amount of time the Captain, the executive officer, or each department head can devote to a certain subject or task represents a valuable investment. A junior officer's time is valuable too. This book will enable you to relate information about specific propulsion equipments and procedures used on board your ship to a *systems concept* which should serve as a beginning for everyone. The OOD must know how systems within the propulsion plant are related. Still, engineering is only one of many subjects with which the OOD must be familiar, and the other subjects require an investment in time and study also.

Understanding just how a propulsion plant operates is difficult for many persons who have made "going to sea" their business. For prospective OODs—particularly those who have yet to set foot on board ship—knowledge of how a marine propulsion plant operates can be a murky subject. Accordingly, this book will use "everyday" language, diagrams, examples, analogies, and even—now and then—a "sea story" to aid comprehension.

Chapter presentations are as follows:

Chapter 1 illustrates how natural laws support the main steam cycle and describes equipments and processes within the steam cycle. Other material in this chapter includes a brief discussion of equipments that support, and are supported by, the main steam cycle.

Chapter 2 discusses "plant insurance policies" utilized to enhance the efficiency and reliability of marine steam plants, which power more than three-fourths of the surface warships of the U. S. Navy.

Chapter 3 concerns the fireroom, or the area where steam is generated which will later be used to do work. Included in this chapter are a description of boiler-related casualties and an illustration of how each can affect ship propulsion.

In Chapter 4 the reader enters the engineroom for some basic information concerning turbines, reduction gears, and lubricating oil systems. Included in this chapter are a discussion of main engine-related casualties and an illustration of how each can affect ship propulsion.

Chapter 5 concerns the men, principally the engineer officers of the watch (EOOW), who exercise control over the entire engineering plant and who serve as important assistants to the OOD. This chapter also contains discussions of engineering conditions of readiness, plant responsiveness, and economical operation of the propulsion plant.

Chapter 6 deals with the information needed by the OOD inport concerning his engineering plant. This chapter explains what is meant by the term "cold iron," among other topics, and explains how an OOD may suddenly find himself "in hot water" during a "cold-iron" period.

Chapter 7 discusses electrical generation and distribution on board a typical warship, as well as the production and control of water for both the propulsion plant and men on board ship.

Chapter 8 concerns gas turbine propulsion, the means of propulsion that will be used by one-third of the destroyer and escort type ships of the U. S. Navy by 1985.

Chapter 9 will be useful to a junior officer assisting in a propulsion plant zone inspection. Included in this final chapter are additional discussions of the entire engineering plant in terms of efficiency and economy.

A secondary objective of this book is to promote confidence in the reliability, responsiveness, ruggedness, and efficiency of the modern marine propulsion plant. A warship or support ship is in reality a combat system, and a reliable engineering plant is a key element of this combat system.

Reliability of the engineering may be viewed differently by different men on board ship: The Captain is confident that his ship will be able to perform all mobility missions, whether in combat or in peacetime. The engineer officer similarly knows that his plant will be "ready to answer all bells." The weapons/combat systems officer assumes that he will have stable electrical power available to his weapons systems, and that this power will enable him to fire under all situations, including the casualty mode. The operations officer is assured that sufficient stable electrical power exists for his search radars and communications gear, and the reliability of the engineering plant enables him to coordinate combat operations for the commanding officer with confidence.

The final objective of this book is to emphasize the desirability of practicing economy in propulsion plant operation. Economy often equates with the phrase "sound engineering practice," a term that crops up often in numerous operating and technical manuals. A brief historical note serves to stress the necessity of economy in propulsion plant operations as both a "sound engineering practice" and as a matter of increasing importance:

In the early 1970s the decision was made to convert nonnuclear ships of the fleet from residual fossil fuel to distillate fossil fuels, such as Navy Distillate (ND), JP-5, and Marine Diesel (DFM). Greater fuel compatibility between different ship types was therefore achieved; however, this "gain" was balanced by "losses" in other areas. In the case of endurance, conversion from Navy Special Fuel Oil (NSFO), a residual fuel, to ND *decreased* the steaming radius of converted ships by about 7%.

In addition to the reduction in steaming radius, conversion of the fleet to distillate fuels meant the Navy would also pay more for its fuel. Conversion began in 1972, when a barrel of NSFO cost about $3.30 compared to $4.50 for each barrel of ND. Significantly, this conversion was begun before the first global energy distribution crisis occurred. Within two years a shortage of Mideast oil and additional refining costs had driven the price of ND beyond $14 a barrel, and since then the price of fuel has generally been tied to the rate of inflation in the national economy.

Even before the fuel price increase experienced engineers always stressed the importance of economical operation on board naval ships. One reason is that propulsion plants that operate economically at 15 or 20 knots, for example, gulp fuel at 27 knots. More equipment must be operated to support higher steaming rates. In addition, that equipment must also be run faster to produce greater volumes and pressures. At even higher firing rates the possibility of overfiring a boiler also occurs if the upper limits of combustion are exceeded. Water consumption increases. Over longer periods of time certain types of metal fatigue can develop in even the strongest of steels.

In most surface ship propulsion installations, increasing to speeds that approach full power also means the number of watchstanders must be increased. Like the amount of fuel a ship carries, the number of man-hours that the engineering force can devote to all of its many tasks is not infinite. But, more about these subjects in later chapters.

To summarize, possessing a basic knowledge of his marine propulsion installation is important to the OOD or a prospective officer of the deck. A competent and confident OOD understands how his plant operates. He appreciates its responsiveness and tries to enhance its reliability by practicing economy wherever possible.

The OOD knows that these subjects are related. Here's an example that is not imaginary:

You are a junior officer serving on board a cruiser. Your ship has been at sea for four days; presently you are two days away from your first port visit—and your ship's refueling stop—during the initial month of your ship's deployment.

Then a highly unexpected message disrupts your ship's routine transit: "INTERCEPT AND TAKE SURVEILLANCE FOREIGN CRUISER GROUP."

Intelligence publications indicate that the foreign cruiser is one of that nation's many modern warships, slightly smaller in size than your ship. Her outline bristles with missiles. According to the message just received, two guided missile destroyers and submarines are in company with the cruiser. And—although you're not too interested in support ships right now—a foreign tanker steams in company with the cruiser group.

On board your ship events in the engineering department happen immediately: The Captain wants the two standby boilers "boosted" to enhance responsiveness and to be ready if additional speed is needed. The fireroom watches respond to this requirement by bringing each boiler to line pressure and subsequently securing each. The engineer officer has already given the chief boiler technician "overload" sprayer plates for assembly in spare burner barrels if additional speed and power are required. (The engineers don't expect to use the "overloads." But they may have to. Everyone wants to be ready.) Meanwhile, standby equipment is rechecked for proper lineup.

The intercept occurs during the early morning—seemingly a surprise to the foreign cruiser and her escorts. The cruiser increases speed and turns away. Her destroyers race into apparent blocking positions. But they are too slow as your ship's powerful engines respond to the Captain's skillful commands.

And then suddenly you're where you want to be—close enough to the cruiser so you can accelerate with her, distant enough so her sudden slowing will not endanger your ship.

As the morning passes, it becomes apparent that the presence of your ship is more than a mere nuisance to the Admiral on board the foreign cruiser. Periodically the cruiser increases speed, and her destroyers stationed abeam close with pincerlike coordination. The tonal burst of the cruiser's cryptographic tactical radio circuit, which seems to precede most evasion attempts, keeps you nervous and alert.

But the cruiser appears to be having minor engineering problems, according to the junior officer of the deck. He is the sonar maintenance officer (and an outstanding shiphandler), but he knows enough about engineering to understand that black smoke from the cruiser means her engineers are having problems with proper fuel combustion. You discover that each puff of smoke signals an impending speed change—the cruiser wants to escape from your ship. But the cruiser cannot, as she continues to "telegraph" her moves like a third-rate boxer. Meanwhile, the stacks of your ship stay clear, from ahead flank to backing bells.

By noon of the second day, the cruiser has seemingly given up trying to escape from your surveillance. But the hard maneuvering has affected your ship. The engineer officer has directed that fuel soundings be increased to twice daily. The oil king's figures show that your ship has used nearly twice as much fuel since the interception was made as was burned in a previous two-day period of routine transit.

Your knowledge of your ship's fuel consumption creates anxiety as you notice the foreign cruiser signal her tanker to take a refueling course. The foreign warships (except the nuclear submarine) refuel. And then, surprisingly, the foreign cruiser group anchors.

Anchoring gives their sailors a rest. More important, perhaps for now, steaming at anchor conserves the foreign warships' fuel. But you can't anchor—what happens if the cruiser weighs anchor suddenly under the cover of darkness? What happens if their nuclear submarine tries to evade?

So your ship remains under way, circling the cruiser group while it remains at anchor.

On board your ship the Captain, the OOD, the engineer officer, and his primary assistants confer. You must conserve fuel. So the Captain gives permission to reduce temporarily engineering responsiveness, but not readiness. The OOD outlines the tactical situation to his EOOW in main control.

With knowledge of what the Captain and the OOD expect, the EOOW directs that both firerooms steam with just one forced draft blower providing air for combustion to the boiler in operation in each space. He directs that electrical rather than turbine-driven pumps be placed on the line to conserve steam. He reviews machinery operating logs in each engineering space to ensure that pressures and temperatures throughout the engineering plant correspond to those that provide the most efficient and economical operation possible. He impresses upon the watch supervisors in each space the importance of saving every possible gallon of fuel.

In the morning the sonar maintenance officer, again on watch as the junior OOD, receives reports that the foreign ships appear to be getting under way. He quickly passes this information to the OOD and the EOOW. Within minutes the engineering plant is ready to respond at its maximum speed on two boilers. It can, in fact, respond at full performance on four boilers in a very short time, since two hours previously the OOD had directed the EOOW once again to "boost" the standby boilers.

Then, suddenly, a new development: As one of the guided missile destroyers hauls around on her anchor to short stay, flames shoot from her forward stack. The brightness of the flames pierces the early morning twilight. Still, no explosion is heard.

The OOD in your ship has only a basic engineering knowledge, but he knows such a fire can occur from excessive carbon collecting in the upper areas of a boiler and in its smokestack as a result of improper combustion. He suspects that this is what has happened in the enemy warship. He surmises that that guided missile destroyer has suffered a boiler casualty which will limit her speed. The Captain comments that the reduced speed capability of that warship may place limitations on coordinated actions by the foreign cruiser group.

Now a certain "routine" sets in as the days pass. The foreign cruiser still makes an occasional attempt to evade, but the attempts appear halfhearted. Perhaps her captain lacks confidence in the engineering plants of his ship and her escorts. But your Captain and the entire crew lack no confidence in the mobility capabilities of your ship.

You know the foreign warships cannot outrun you. But now you begin to worry that they may be able to outlast you as fuel consumption becomes the primary concern. The foreign warships refuel regularly. Your fuel oil storage tanks are being filled also—only the liquid in many of your storage tanks is salt water as the oil king and his assistant begin to take on ballast.

On board your warship daily fuel consumption stays constant, however. More important, consumption remains low. The engineering measures taken by the Captain, the OOD, and the engineers as a team are paying dividends.

On board the foreign cruiser the admiral's staff uses its intelligence publications to estimate the capabilities of your ship to continue her surveillance mission. Perhaps they think your ship now burns fumes instead of fuel oil! They are wrong—you don't have much fuel left, but you can still perform your mission for at least a few more days.

Then comes the message you've been awaiting: A friendly oiler is nearby. She has fuel for you, and destroyers ready to take over your surveillance duties are not far behind.

Once again, your ship has performed her mission. The performance of your engineers and engineering plant has been a major factor in this success.

Information in the chapters that follow will describe how this success becomes a matter of routine.

Acknowledgments

Many men contributed to this book. They range from admirals and captains, chief petty officers and civilian engineers, to an elementary school teacher in Norfolk, Virginia, whose name I'll probably never know.

The third grade teacher is included because he seemed to understand, just about as well as anyone ever can, the importance of a sound engineering plant to a warship. This may be your conclusion too, since that teacher assigned an A+ for the following essay, written by the son of a commanding officer of a *Charles Adams*–class guided missile destroyer:

My Three Wishes

> First of all, I wish for peace for the world. My second wish is for my family to always be together and be happy. My last wish is that my Dad's engineering plant gets well and stays healthy so my Dad can spend some time thinking about the rest of his ship.

On a more serious note, the subject of shipboard engineering involves many more people than the nine-year-old children of the men who man our surface warships. I would like to thank two organizations and acknowledge the particular involvement of nine men whose ideas contributed to this book.

The first organization is the U. S. Naval Institute, for access to information and illustrations from some of its publications, which I have included in this book. The second organization is the Naval Sea Systems Command, Destroyer DD 963 Ship Acquisition Project (PMS 389). PMS 389 reviewed the material contained in Chapter 8 that pertains to *Spruance*-class destroyers and similar engineering systems of *Perry*-class guided missile frigates.

There are three senior officers in particular whom I would like to thank for their insights on "what the Captain expects his officers of the deck to know about their engineering plant." Rear Admiral Robert M. Collins, USN, and Captain Robert E. McCabe, USN (Retired), were the truly superlative and inspirational commanding officers of the USS *Fox* (CG 33) when I served as engineer officer of that ship. Captain William B. Latham, USN, was the first senior member of the Propulsion Examining Board, U. S. Atlantic Fleet, whose most recent at-sea tours have included command of a destroyer squadron and serving as chief of staff of an aircraft carrier group.

They, as well as Lieutenant Commander Thomas F. Madden, USN, who reviewed the material on gas turbine propulsion contained in Chapter 8, are only a few of the many officers in the Navy vitally concerned with excellence in the engineering plants of our surface warships.

Finally, there are five men whose suggestions, more than those of any others, helped "write" this book. All are now retired master chief petty officers of the U. S. Navy: Machinist's Mates Jack Price and Earl Hackley, Boiler Technicians William Rathbone and John Hogberg, and Electrician's Mate Jimmie Edmisten. Each of these men, former teammates on the Propulsion Mobile Training Team (Norfolk) of the Naval Surface Force, U. S. Atlantic Fleet, worked long hours to upgrade the engineering expertise of all officers and men with whom he served.

ENGINEERING FOR THE OFFICER OF THE DECK

1

What Is Supposed to Happen, and Why It Does

Remember the first time you stepped into a main machinery space when the engineering plant was in operation? If your thoughts were similar to those of many of your contemporaries, the experience may have been something like this: "Those chief petty officers never told me it would be like this! All these strange noises, the whinings and the hissings—how can anyone hear anyone else? And what about all these gauges and indicators? Seems like there are so many that nobody knows which ones to read. . . ."

But after a while you did become acclimated, even though you didn't think you would. The whine of a turbine-driven pump no longer sounded quite so threatening. Looking up and about, you realized that those steam lines you had traced in a classroom as straight red lines on clean white paper really do need to be curved to allow for thermal expansion. The odors of hot coffee and hot lube oil mingled in a pleasing aroma. The men on watch about you obviously knew their jobs, and they told you that the operation of the propulsion plant certainly didn't seem like a mystery to them.

Then—just about the time you were feeling relaxed—a nearby air compressor set for automatic starting switched on with a loud clatter. You didn't jump. But you quickly glanced toward the ladder that would take you up and out of the machinery room if you had to leave in a hurry. You hoped no one noticed. Someone did, but his smile only meant that he too had had a similar experience. . . .

Certainly the sights, sounds, and even smells of a shipboard propulsion plant can be mysterious when encountered for the first time. Some men claim that "what's down there" is a mystery on even their fourth or fifth excursion into an engineroom or a fireroom. But there are no actual mysteries surrounding the operation itself of a properly maintained propulsion plant. All of the machinery and equipment in the engineering plant is designed to change energy from one form to another or to transport energy from one place to another. In doing so, the engineering plant operates according to natural laws. Put another way, those water molecules and speeding electrons racing through the various systems of a conventional propulsion plant *must* respond to natural laws—as long as men strive to operate their plant according to these laws.

The First and Second Laws of Thermodynamics govern the operation of propulsion plants. This chapter will also deal with topics that are important to achieving a basic understanding of the marine propulsion plant—the main steam cycle, the pressure/temperature relationship, superheat and fuel, and the speed/power/fuel relationship, to name just a few. But understanding how a conventional propulsion plant operates begins with the first two laws of thermodynamics.

THE BASIC PROPULSION PLANT

Briefly, the First and Second Laws of Thermodynamics may be paraphrased as follows:

First Law—Many processes, such as the chemical transformation involved when fuel burns, can produce heat or do work.

Second Law—Thermal energy flows from a warmer to a cooler region.

Now, taking the simplified definitions stated above and adding an earlier statement that all the machinery and equipment in the plant is designed either to change energy from one form to another or to transport energy from one place to another, it's time for a look at a simplified marine propulsion plant. The plant in Figure 1-1 would be a tremendously inefficient propulsion plant and serves as a demonstration only, but it illustrates the fundamentals of plant operation quite clearly.

The process begins in the boiler firebox, where chemical energy contained in fuel oil is transformed into thermal energy during the process of combustion. Thermal energy flows from the intensely hot firebox to the relatively cool feedwater, and steam is generated. The increased temperature and pressure of the steam generated inside the boiler indicate that thermal energy is being stored internally within the steam; this thermal energy contained in the steam will be converted to mechanical energy. Mechanical energy will produce work as the turbine wheel moves, so that ultimately the shaft will turn, and the ship will move through the water.

Using increased pressure as a vehicle and the physical confinement of the main steam line as a highway, the thermal energy contained in the steam rushes toward the turbine. Two main energy changes must take place before thermal energy can be converted to work by the turbine. The first change occurs when the thermal energy contained in the steam is

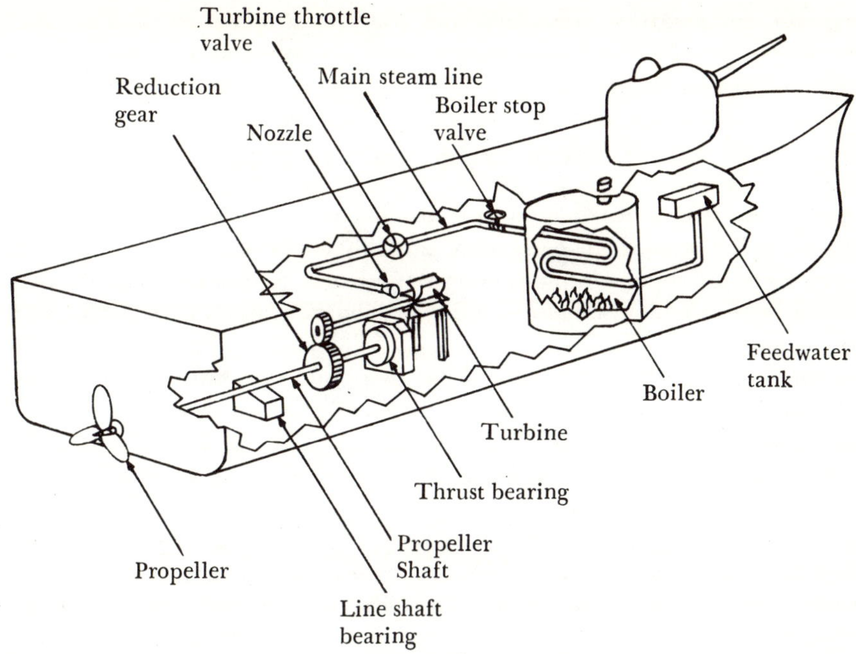

Figure 1–1. Basic ship propulsion

transformed into mechanical energy as the steam passes through a nozzle. The second energy change takes place instantaneously when the mechanical energy that the steam now possesses impinges upon the turbine blades, and the turbine begins to turn.

The turbine, a rotary engine consisting of curved blades mounted on a shaft, turns either from reaction to the steam impinging upon, or by impulse as the steam impinges upon the blades, or through both impulse and reaction, depending upon design. The turbine is connected to the propeller shaft by a reduction gear. The reduction gear matches the relatively high rotational speed of the turbine to the lower speed of the propeller, thereby allowing each to operate within its respective efficiency range.

As the propeller revolves, it pushes against the water. The water pushes back with equal force. The thrust produced by the water pushing back is transmitted along the propeller shaft to the thrust bearing. The thrust bearing permits rotation of the propeller shaft, while the bearing simultaneously transmits thrust to the ship's hull. The resulting force drives the ship through the water at a speed proportional to the rotational speed of the propeller.

If a propulsion plant identical to that shown in Figure 1-1 really existed, it would be a model of *in*efficiency. Such a plant would be unacceptable for a warship, and the owner of a fleet of merchant ships powered by similar plants would go bankrupt. Such a plant would waste a tremendous amount of energy.

PROPULSION EFFICIENCY

Any propulsion plant is designed around an energy balance. Figure 1-2 shows such a balance. The diagram shows that some amount of energy loss is inevitable. Propulsion plant designers attempt to minimize energy losses through precise plant design in order to enhance efficiency. On board ship, plant operators strive to limit energy losses by following specified maintenance procedures and by operating their plant according to exact operating standards. These plant operators would tell you that the basic propulsion plant shown in Figure 1-1 is inefficient for the following reasons: (1) thermal energy which turns feedwater to steam is lost to the system, since the steam/water fluid is not recovered after expansion through the turbine; (2) more thermal energy is lost up the stack, since the feed water makes only a very few passes through the boiler; and (3) even more thermal energy is lost by radiation through the main steam line, since this line is not insulated (engineers would use the term "lagged") to prevent this energy loss. The plant operators would add that

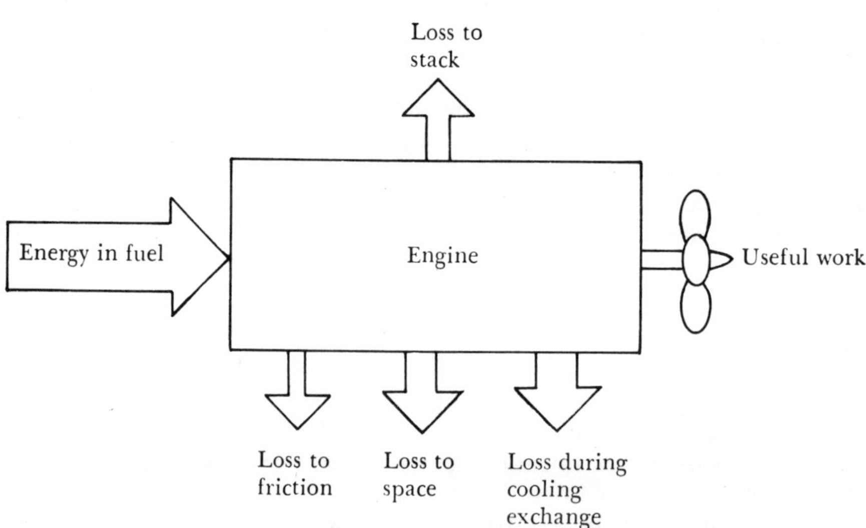

Figure 1–2. The energy balance

the subject of energy loss through friction (to be covered in Chapter 4) is also important. Plant designers and operators alike would conclude by saying that early propulsion systems were similar to that shown in Figure 1-1, since relatively large amounts of fuel had to be burned in those inefficient plants to compensate for the large energy losses illustrated in Figure 1-2.

The requirements of a modern warship's propulsion plant include reliability, ruggedness, and responsiveness. As weapons systems/electronic suites become larger and more sophisticated, a low space/weight to a high power output ratio becomes highly desirable as a military feature of naval propulsion plants. The use of petroleum fuels and numerous design improvements made since the U. S. Navy began its transition to steam in the mid–nineteenth century account to no small degree for the increased efficiency of the conventional steam propulsion plant. However, the development of stronger steels and high-temperature alloys—also a major factor in the development of the gas turbine and other power plants—accounts for much of the improved efficiency of the modern steam propulsion plant.

Theoretically, the work that can be obtained from a steam turbine equals the total energy change from inlet to exhaust (outlet). (The actual amount of work produced by a steam turbine depends upon the efficiency of that turbine in converting thermal energy to mechanical energy.) Turbine installations in marine propulsion plants exhaust into a condenser, which condenses the remaining steam as that steam expands through the turbine and comes in contact with seawater circulating in tubes through the condenser. The temperature at which this condensation takes place depends greatly upon the temperature of the surrounding seawater. Therefore, for any turbine in a static environment, efficiency will be increased when the amount of thermal energy available to the turbine is increased.

Increasing the energy available to a steam turbine can be brought about either by increasing the quantity of steam flowing to the turbine or by increasing the thermal energy contained in the steam. The first method requires a larger boiler and condenser, as well as a probable increase in the size of supporting equipments such as feed and condensate pumps. The second method is more desirable in ship construction, since this method is compatible with a decrease in the size and weight of propulsion machinery.

Use of the improved steels and alloys in the main pressure parts and equipments of the modern propulsion plant enables steam at higher temperatures to be used. The propulsion plant of a typical U. S. destroyer of the 1930s used steam at 565 psi (lb/in^2) at 700°F. Destroyers and cruisers that began to enter the fleet in large numbers in the late 1950s and the 1960s use steam at 1200 psi and 950°F. The increase in temperature is important because for every 35°F increase a naval steam propulsion turbine gains about 1% in efficiency. Use of this highly "superheated" steam in modern engineering plants means, in practical terms, that we can put more weapons in a ship without increasing the size of the ship at the expense of its propulsion plant.

FUEL AND SUPERHEAT

A warship or any other ship carries only a certain amount of fuel. In addition to the limited amount of fuel, the commanding officer and the engineers of his ship must contend with the fact that only a certain amount of energy is available from that fuel. All of that energy must be used efficiently.

The amount of energy available in a given measure of fuel varies according to the type of fuel. For example, a gallon of distillate fuel similar to Marine Diesel (DFM) contains approximately 139,000 British thermal units (BTU), while a gallon of residual fuel similar to Navy Special Fuel Oil (NSFO) contains approximately 148,000 BTU. (One British thermal unit is defined as the quantity of "heat," also referred to as thermal energy in transition, required to raise the temperature of one pound of water one degree Fahrenheit. Both the calorie, a unit of measurement used with the metric system, and the BTU can be expressed in terms of *joules;* this provides a means of converting measurements of thermal energy to units of work.) While the number of BTUs in a gallon of any petroleum fuel sounds large, it is important to remember that there are also large energy losses in a power plant, as shown by Figure 1-2. Conserving the energy of all those BTUs and making the most efficient use of that energy are challenges that face both the designers and the operators of naval propulsion plants. The efficiency of a conventional propulsion plant is increased by superheating the steam to be used by the turbine.

Engineers who operate the propulsion plant of your ship or any other ship are constantly concerned with maintaining key pressure-temperature relationships throughout the plant. These temperatures and pressures are indications of the energy that exists at particular points of the propulsion system at any given instant. An appreciation of the temperature-pressure relationship is also important to understand how steam may be superheated.

When a liquid boils, a vapor is generated. As long as the vapor remains in contact with the liquid from which it is being generated, both vapor and liquid will have the same temperature. In this condition the vapor and liquid are in *equilibrium contact* with each other, and the temperature at which the boiling liquid and vapor exist in equilibrium contact depends upon the pressure under which the process takes place. As the pressure increases—or decreases—so does the boiling temperature. A liquid and a vapor in equilibrium contact are said to be in a *saturated* condition.

The important thing to remember about the generation of steam (or any other vapor) is that it is impossible to raise the temperature of the steam beyond that of water as long as both remain in a saturated condition in equilibrium contact. This occurs at about 486°F in 600 PSI plants and at about 567°F in 1200 PSI plants.

However, the propulsion turbines in these respective plants are not very efficient at these temperatures, and in the case of the propulsion and electricity-generating turbines in 1200 PSI plants the effect of saturated steam impinging upon turbine blading can be disastrous.

Superheated steam that will satisfy the design requirements of propulsion plants can be created by removing the steam from contact with the water and adding more thermal energy to the steam. This dry steam picks up the additional thermal energy in the superheater of a propulsion boiler. The increased efficiency resulting from the use of superheated steam reduces the amount of fuel oil required to generate each pound of steam and creates size and weight savings throughout the propulsion plant. These are the advantages of using dry, superheated steam:

1. Overall plant efficiency increases, thereby creating savings in fuel and in space and weight.
2. Superheated steam causes relatively little corrosion and erosion of machinery and piping.
3. Superheated steam loses its additional thermal energy relatively slowly.

SPEED, POWER, AND FUEL

The preceding section discussed the importance of superheat in conventional steam propulsion plants for warships as a means of achieving greater fuel economy. Warships such as destroyers are required to operate over greater speed ranges than are merchant ships. Efficient operation is desired from the propulsion plants of all ships. Of major importance to the OOD of a warship is the knowledge that while his propulsion plant may operate efficiently at high speeds, the plant will not do so economically.

The power output of a prime mover such as a naval propulsion turbine is measured in terms of horsepower, with 1 HP = 33,000 foot-pounds of work/minute. Different types of prime movers are rated in different kinds of horsepower. Steam turbines are rated in terms of shaft horsepower (SHP), which can be measured by a torsionmeter mounted on a propeller shaft. Gas turbines used for propulsion, which will be discussed in Chapter 8, are rated in terms of brake horsepower (BHP), which is measured at the point where the turbine clutches to its associated reduction gear. (In some gas turbine propulsion plants, shaft horsepower is also measured; SHP could be used to limit gas turbine speed in those installations in which propeller shaft construction does not have the ruggedness to match the work output capability of the gas turbine.)

Propulsion plant designers have determined that the following relationships exist:

1. The speed of the ship varies with shaft revolutions per minute (RPM) in a nearly linear relationship.

2. Power varies approximately as the cube of propeller speed, which is dependent upon shaft RPM. (Designers use more exacting formulas in actual ship design.)

3. Fuel consumption varies directly with power.

The *Principles of Naval Engineering,* Chapter 5, describes the relationship between speed and power in concise terms for general line officers: "Since the power required to drive a ship is approximately proportional to the cube of propeller speed, 50% of full power will drive the ship at about 79.4% of the maximum speed attainable when full power is used for propulsion, and only 12.5% of full power is needed for 50% of maximum speed." (This statement refers to a ship operating under steady running conditions.)

When fuel consumption becomes a factor in the equation concerning speed and power, elements of the preceding statement can be used to express this relationship. Simply stated, at the higher reaches of shaft horsepower the ship burns a tremendously large amount of fuel without producing a correspondingly greater increase in speed. A warship such as a destroyer or cruiser can make high speeds, but it cannot do so economically.

ECONOMICS AND STEAM PROPULSION

Throughout this chapter we have very briefly discussed the conventional plant in terms of efficiency and economy. In following sections we will see how equipment duplication enhances reliability, and how the use of steam at decreased pressures and temperatures is used to support propulsion. (Other topics will also be examined.)

The modern conventional steam plant which powers the majority of warships of the U. S. Navy and warships of many other navies is reliable, efficient, and economical. The following section will briefly describe this economical propulsion system. Before we encounter the basic steam cycle, however, it may be helpful to relate the conventional steam plant to some aspects of modern-day economics. Consider, for example, this analogy:

- The boiler (heat source) is a manufacturing plant.
- The product the boiler manufactures for sale is steam.

- The buyers consist of the main engines, turbogenerators, and (in most ships) pumps to support the propulsion process.
- Almost all the product (steam) is consumed.
- A raw material (feedwater) is created by natural processes and returned to the manufacturer.
- The rate of exchange depends on various pressure-temperature relationships existing in the "economic" cycle.

Now—with the understanding that we have taken only a very quick and greatly simplified look at marine propulsion system design (and an even briefer look at the subject of modern-day economics!)—we will consider the basic steam cycle employed in the conventional steam propulsion plant.

BASIC STEAM CYCLE—THE PROPULSION PLANT SUPERSTAR

The preceding section discussed the conventional propulsion plant in terms of contemporary economics. While this may be a useful analogy for illustrating the importance of major equipments, little has been said about the role of the steam cycle itself in the conventional propulsion plant.

Boiler technicians, who speak of their boilers in glowing terms, and machinist mates, who are equally enthusiastic about their main engines, agree that the real "superstar" of the "propulsion production" is the steam cycle. Think of the boiler, or the main engine, or even a condensate pump or a main feed pump, for that matter, as a highly important but nevertheless supporting actor. Each plays an important role, but maximum performance by any one will not guarantee success of the production. Success depends upon a healthy steam cycle.

Knowledge of the basic steam cycle is important to the OOD because the steam cycle truly is the superstar of the propulsion production. The steam cycle depends upon top performance from the boiler, engine, condenser, and various pumps and other equipments to deliver its own maximum performance. The role of the steam cycle is so important that a "supporting cast" is also required just to ensure that the performance of the steam cycle does not falter. This supporting cast includes the following: auxiliary steam systems, decreased pressure auxiliary steam systems, fuel oil service system, electrical system, and equipment duplication. With the exception of the fuel oil service system (included in Chapter 3) these topics will be discussed briefly in this chapter.

Figure 1-3 shows the four phases of the basic steam cycle.

Generation Phase

The generation phase begins after relatively pure water that has been freed of entrapped air is pumped into the boiler under pressure. Thermal

Note: Economizers are not associated with the 1200 PSI pressure-fired steam generators of the FF-1040 and FFG-1 class warships.

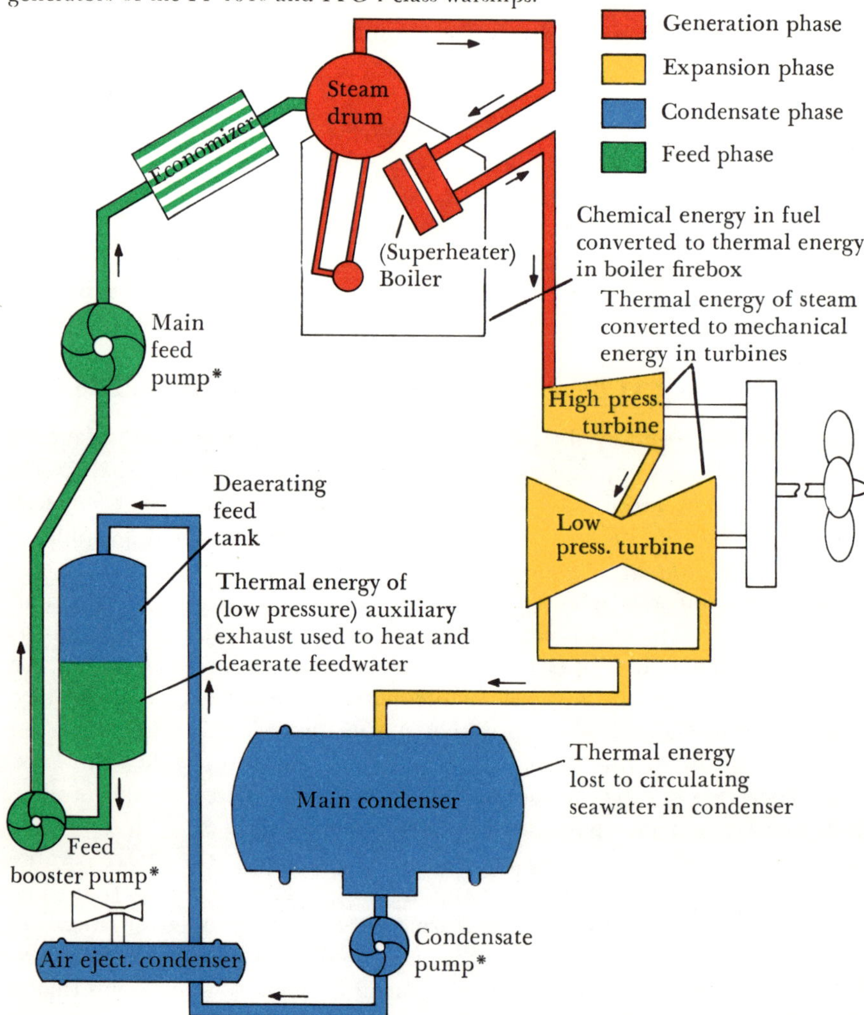

* Conversion of energy occurs at this point, with the respective pump converting thermal energy (turbine-driven pump) or electrical energy (motor-driven pump) to mechanical energy in order to move a fluid *toward* a region of increased heat.

Figure 1–3. The basic steam cycle

energy released during combustion flows through *generating tube* walls to this water, and steam forms as the water circulates through the generating tubes. This saturated steam collects in the *steam drum,* where *drum internals* remove much of the moisture from the steam before it passes through a *dry pipe* (termed a dry box in some boilers) and into the *superheater.*

The superheater consists of additional tubes whose surfaces expose the now-dry steam to even more thermal energy. After emerging from the superheater, superheated steam flows through an outlet valve known as the *boiler main steam stop* (not shown) and rushes toward the main engine through the *main steam line.*

Expansion Phase

In the expansion phase, the thermal energy of the steam will be converted into mechanical energy, which will result in work being accomplished. This phase begins when the steam passes through the *throttle valve* (not shown) and encounters the first set of turbine nozzles. The *high pressure (HP) turbine,* a rotary engine consisting of progressively larger stages of *nozzles* and *blades* mounted upon a central shaft, first uses the steam while it has a relatively small volume but high temperatures and pressures. The *low pressure (LP) turbine,* a rotary engine consisting of progressively larger stages of both fixed and moving blades, receives the steam next. Located downstream and larger in size than the HP turbine, the LP turbine extracts additional energy from the steam before exhausting it into the main condenser. Use of an HP and LP turbine combination is highly efficient because the LP turbine removes a large additional amount of energy from the steam before it enters the main condenser, which is at a pressure below atmospheric pressure (14.7 psi at sea level). (Some turbine installations also include a cruising turbine, although not those of the more modern steam-powered cruisers and destroyers.) For astern operation, the steam is diverted so that its only path of flow is to the *astern elements,* which are located one on either side of the LP turbine. (These astern elements are mounted with the LP turbine ahead elements and contained in the LP turbine casing on a common rotor shaft; however, steam is not admitted to both the ahead and the astern elements simultaneously.)

Condensate Phase

The condensate phase begins when steam exhausts from the final stages of the LP turbine into the *main condenser.* The main condenser is the relatively cool region toward which energy flows from the heat source (boiler). The main condenser also provides the means for recovering water for subsequent return to the boiler as feedwater.

The main condenser, a massive heat exchanger suspended from the bottom of the LP turbine in most propulsion installations, is simultaneously an equipment that increases propulsion efficiency and the scene of a major energy loss. A vacuum, initially created and later sustained by the *air ejector,* occurs when relatively cool seawater circulating in tubes through the condenser rapidly condenses steam exhausting from the LP turbine. This vacuum increases the temperature difference between the heat source and the receiver, thereby increasing the thermodynamic efficiency of the main engine.

The condensation process also enables water to be recycled. A single propulsion boiler can generate more than 150,000 pounds of steam in one hour, and shipboard distilling plants cannot supply this demand. The main condenser performs a vital function in recovering water to support the steam cycle, but in this process a tremendous energy loss occurs when the steam surrenders its thermal energy to the cooler seawater during the heat exchange. (Figure 1-2 illustrates this loss.) Even in modern steam propulsion plants, as much as 60% of the total energy input is lost during condensation. However, this loss must be compared to the even greater amount of energy that would be required to distill constantly water for steam generation in a propulsion plant with no condenser.

The water that has condensed and fallen to the bottom of the main condenser as condensate will eventually return to the boiler as feedwater. However, we are now concerned with moving this water from a cool region toward warmer regions and finally into the boiler itself. From the point at which condensate collects in the *hot well,* pumps must be employed to do this work. The first pump required is the *condensate pump,* which develops sufficient pressure to remove water from the hot well and push water through the *air ejector* condenser to the spray nozzles inside the *deaerating feed tank* (DFT).

Feed Phase

The feed phase begins when condensate emerges from the *spray nozzles* as a fine mist inside the dome of the DFT. This mist surrenders its entrapped air and collects in the bottom of the DFT as feedwater. Meanwhile, the air released in the process is drawn from the top of the DFT. The *main feed booster pump* takes a suction on the bottom of the DFT and discharges the feedwater to the *main feed pump.* The main feed pump pushes the feedwater at a greatly increased pressure through the boiler *economizer,* where additional thermal energy is added to the energy already gained in the DFT as firebox combustion gases pass between the economizer tubes. The heated feedwater is discharged to the steam drum at a pressure sufficient to ensure that drum water level is maintained at a level which will support all steaming demands.

Generation, expansion, condensate, and *feed*—nothing is mysterious about the basic steam cycle. However, when you first begin systems tracing —physically sighting condensate and feedwater lines, as well as easily identified equipments such as the boiler, turbines, and condensers—the maze of piping and supporting equipments can be intimidating. So here's how one chief petty officer, an excellent engineer officer of the watch (EOOW), explains the basic steam cycle to men who are new to his propulsion plant:

"First of all, forget those pipes for now. You'll get acquainted with piping and supporting equipment later. Just think of the boiler as the heat source and the condenser as a cold sink. Energy is going to go from the boiler, through the turbines, and to the condenser, just like the thermodynamics law says it must. Then after the condenser, you've got to be pumping to get water back into the boiler."

The chief might have made the point that pumps convert either thermal energy or electrical energy, depending on whether the pump is turbine-driven (T) or motor-driven (M), into mechanical energy in order to move a fluid. Since the combustion process in the boiler converts chemical energy into thermal energy, and thermal energy flows from a warmer to a cooler area, the chief's illustration (Figure 1-4) expresses the steam cycle in terms of a simplified energy cycle.

Figure 1-4 illustrates the importance of pumps to the basic steam cycle. Although not components of the basic steam cycle, the forced draft blower —a giant air pump supplying air for combustion—and the fuel oil service pump must deliver the proper amounts of air and fuel to support the generation phase. In the condensate phase if the condensate pump malfunctions, condenser vacuum, air ejector operation, and the DFT water level will all be affected. The feed phase depends upon sufficient hot, deaerated water being in the DFT to supply a positive pressure to the main feed booster pump. So if the condensate pump has malfunctioned, the feed phase is already in jeopardy. Generation, of course, depends upon sufficient feedwater being delivered to the boiler to support all steaming demands.

Pumps are so important in supporting the basic steam cycle, the propulsion plant's superstar, that they are duplicated in the condensate and feed phase to ensure that water continues to move toward the boiler. Duplication also occurs in the auxiliary steam systems, the electrical system, and the fuel oil service system. All of these systems support the basic steam cycle. With the exception of the fuel oil service system (discussed in Chapter 3), the relationship of each of these systems to the basic steam cycle will be discussed in this chapter.

EQUIPMENT DUPLICATION

Equipment duplication, also termed equipment redundancy, in all naval propulsion plants gives warships propulsion reliability. In conventional steam propulsion plants the following is true:

- Each main engine can be supplied with steam from more than one propulsion boiler.

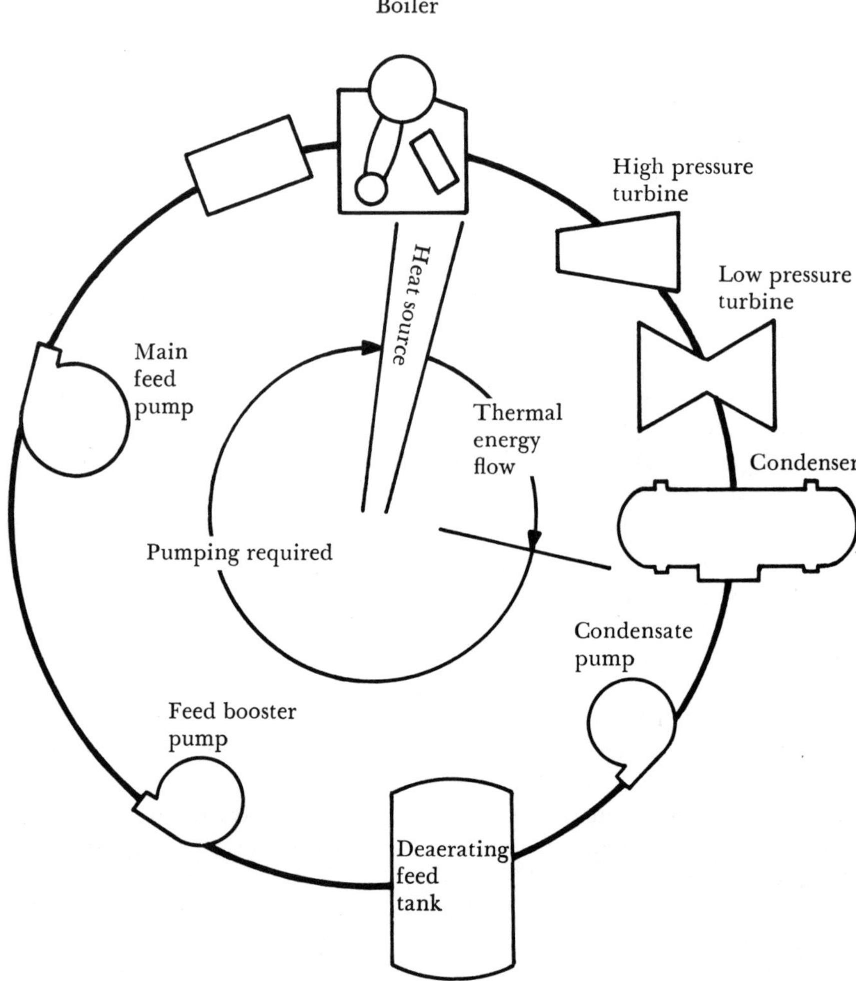

Figure 1–4. The energy cycle

- Every main boiler, except pressure-fired steam generators on board the FF-1040 class and FFG-1 class ships, has at least two forced draft blowers.
- Every propulsion system has at least two fuel oil service pumps.
- Every propulsion system has at least two condensate pumps.
- Every main condenser has two separate air ejectors.
- Every propulsion feedwater system has at least two main feed pumps, with each main feed booster pump (which must supply a main feed pump) capable of discharging to any main feed pump in most installations.

For reasons of fuel economy, ships usually steam with the minimum amount of equipment that will reliably support the mission. However, it is important to remember that equipment duplication helps make the conventional steam plant a highly reliable propulsion plant.

AUXILIARY STEAM SYSTEMS

Auxiliary steam systems supply steam at reduced temperatures and pressures for the operation of many systems and equipments throughout the ship. In the propulsion plant of a 1200 PSI destroyer or cruiser, a plant that powers the majority of our surface warships and will therefore be used in many examples throughout this book, main steam is used by the main engines, main feed pumps, and turbogenerators. Other equipments throughout the plant *could* use this 1200-psi, 950°F steam, but the cost of using the special alloys and metals necessary to handle this high temperature and pressure would be prohibitive. Wherever it is applicable, it is cheaper and more practical to use auxiliary steam which has adequate temperatures and pressures for the task at hand.

In a 1200 PSI ship, auxiliary steam is produced by diverting some of the main steam into a *desuperheater* inside the boiler's steam drum or water (mud) drum. A heat exchange results, with the steam giving up thermal energy to the surrounding water circulating inside the drum. The steam emerges, still at 1200 psi, but with its temperature greatly reduced. (The steam in this state still retains some superheat; 600 and 150 PSI systems also retain small amounts of superheat.) The only equipment that uses this 1200-psi, desuperheated (DSH) steam in 1200 PSI CGs, DD/DDGs, and 1052/78 class FFs is the forced draft blowers.

However, listed below are *all* the uses of 1200-psi DSH steam in both the forward and after plants of a 1200 PSI DD/DDG or CG:

Forced draft blowers
Auxiliary exhaust augmenting station
1200-600 PSI reducing stations (2)
Soot blower header (in some ships)

As indicated above, two 1200-600 PSI reducing stations are in each propulsion plant of a 1200 PSI DD/DDG or CG. Duplication illustrates the importance of the 600 PSI auxiliary steam system. Here are the uses of this system:

Fuel oil service pumps
Main feed booster pumps (T)
Fire and flushing pump (T)
Lube oil service pump (T)
Main condensate pump (T)

Main condenser seawater circulating pump
600-150 PSI reducing stations (3)

(Motor-driven pumps are installed in some of these ships for equipment duplication with turbine-driven pumps. This duplication provides back-up in case steam pressure is lost. Turbine-driven pumps are more reliable than electric pumps, but the latter have the capability of receiving power from an alternate emergency source. As an additional note, the absence of turbine-driven pumps, except main feed pumps, in 1200 PSI propulsion plants in the FF and FFG-1 class ships in favor of electric pumps results in simplification of auxiliary steam systems in those ships.)

Here are the more important uses of 150-psi steam, as far as the OOD is concerned:

Air ejectors
Ship's whistle (forward plant only)
Auxiliary exhaust steam augmenting station
Lubricating (lube) oil heaters

Other pressure and temperature reductions are necessary in the plant to ensure that there is an adequate supply of steam for the ship's distilling units, main engine gland sealing system, galley, and many other important uses. A constant supply of auxiliary steam is important; in considering the basic steam cycle, it is important to remember that main steam cannot be produced without equipments that use auxiliary steam as an energy source.

DECREASED PRESSURE AUXILIARY STEAM SYSTEMS

As illustrated in Figure 1-3, the condensate phase ends, and the feed phase begins, inside the DFT. In addition to these two phases meeting inside the DFT, so do two highly important decreased pressure auxiliary steam systems. The relationship between the auxiliary exhaust system and the high pressure drain system is an interesting one because both systems are by-products of plant operation, yet both systems are required for the plant to operate. Auxiliary exhaust and high pressure drains meet in the DFT, where both are required to help turn condensate into feedwater.

Auxiliary exhaust may be thought of as "leftover" steam. Not all the steam supplied to a turbine-driven pump is used when the pump operates. Spring-loaded exhaust valves in the pump (or forced draft blower) casing permit this "leftover" steam to enter the auxiliary exhaust main at a designated pressure. Although the pressure of auxiliary exhaust is only 15 psi in most plants, there is a considerable amount of thermal energy in auxiliary exhaust. Moreover, the total volume of auxiliary exhaust is quite large.

Auxiliary exhaust performs two important tasks in the engineering plant:

1. It "scrubs" condensate of entrapped oxygen and helps heat the condensate as the condensate emerges from the spray nozzles inside the dome of the DFT.

2. It provides the heat source for the salt water heater in the distilling unit, where the conversion of seawater to usable potable water or feedwater begins.

The thermal energy contained in high pressure drains is retained in the propulsion system by continuously draining all steam lines with a pressure of 150 psi or greater. Steam traps and orifice-configured traps perform this function by draining these lines and passing the resulting water and wet steam into the HP drain line. The HP drain line provides a path for thermal energy to flow into the DFT. HP drains are used to heat the DFT so condensate can be converted to feedwater.

The relationship of HP drains, auxiliary exhaust, and the DFT is interesting and illustrates the interdependence of some of the systems with which the engineers must contend. Here's an example:

A number of HP steam traps are malfunctioning, allowing excessive pressures to develop inside the HP drain main. This in turn causes the DFT shell pressure to rise, which greatly exceeds the pressure in the auxiliary exhaust line. Interesting events begin to occur: Auxiliary exhaust, which can no longer enter the DFT, begins to dump to the main condenser. Condenser vacuum drops off slightly, at least at first. The DFT can no longer accomplish its function, since there is no auxiliary exhaust to scrub the condensate free of entrapped air and oxygen. Feedwater, which now contains more than the maximum permissible amount of dissolved oxygen, begins to enter the boiler.

The EOOW knows that dissolved oxygen in boilerwater (feedwater inside the boiler) at the tremendously high temperature and pressure in any naval propulsion boiler will eventually result in a catastrophic boiler casualty because of corrosion and metal deterioration. However, right now the EOOW is more concerned about vacuum in the main condenser, since in this case vacuum has now decreased to a point where the ship's capability to make high speeds is affected. (This subject will be covered in Chapter 4.)

At this point the EOOW orders the malfunctioning steam traps cut out of the HP drain main. This will solve his immediate problem, except now water and energy, which should properly be returned to the steam cycle, will instead be lost to the bilges. Cold makeup feedwater must be taken on in the main condenser to balance this loss. That's no problem—as long as the ship's distilling units can keep up with the increased demand for makeup feed, and fuel usage is not critical.

Perhaps the EOOW decides to order that HP drains to the DFT be secured completely—now the DFT will lose a valuable heat source. The boiler firing rate must be increased to compensate for the fact that relatively cool feedwater is entering the boiler. Fuel will be wasted. Beyond this, prolonged overfiring of the boiler can result in brickwork failures within the firebox.

In the modern conventional propulsion plant, a seemingly insignificant system such as auxiliary exhaust or HP drains has great importance. But now it's time to let the EOOW in this example solve his own problem and to consider the ship's electrical system.

THE ELECTRICAL SYSTEM

Although Chapter 7 is devoted to the electrical plant, the electrical system is mentioned here because this system is required in most plants to support the basic steam cycle. Ships, in fact, could not function without electricity, and the comparison of a ship to a city dependent upon electricity is valid whether the ship is under way or inport.

Substitute turbogenerators for the turbines in Figure 1-3, and the steam cycle remains the same. Replace the propeller in Figure 1-2 with a quantity of electrical power, and the energy balance diagram remains the same.

This electrical energy is converted to work to perform numerous tasks in the propulsion plant. Electric pumps exist in combination with turbine-driven pumps or have replaced turbine-driven pumps completely in many plants. Ventilation fans are needed to pump air into and out of engineering spaces. Alarms alert engineers to unsafe conditions. Compressors supply air for automatic controls and starting air for internal combustion engines.

Outside the engineering plant, this work is vital to the very reason for a warship's existence: the illumination of enemy aircraft by a DDG's missile control radar; the movement of stores from storage to replenishment stations on the deck of an AFS; the lowering of boats to land the Landing Force by an LKA.

Production of electrical energy begins with the generation of alternating current (AC) electricity by the ship's turbogenerators (SSTG), or the gas turbine (SSGTG) or diesel generators (SSDG) for emergency purposes in steam-powered ships. The higher voltages permissible with AC generators and the utilization of AC motors and other AC equipment result in considerable savings in weight and space required for the electrical installation on board ship.

(All ships require at least some DC electricity, however. AC motors do not adapt well to situations requiring a wide range of speed controls, which is one reason why some DC is required. However, AC can be converted to DC and vice versa to satisfy all requirements of all ships.)

On board ship men control the flow of electricity from one point to another with the power distribution systems. These systems—ship's service system, emergency power system, and casualty power system—are the connecting links between the generators which supply electric energy and the various equipments that convert this energy to work.

Electricity is vital to the propulsion plant, the mission of the ship, and every man on board ship. When electrical power is lost or even interrupted, the situation becomes crucial in the propulsion plant and both crucial and expensive to the ship's weapons systems. Loss of electrical power can also be psychologically devastating. In the words of one OOD: "If the engineers have a casualty, shouts and sounds will at least tell me something is happening, although I hope the word the EOOW passes up through the 1JV talker precedes those sounds. But if the lights suddenly go out, that's when you really get a helpless feeling!"

CHAPTER FOOTNOTE

Those men who make the business of going to sea their business know that there are very few times on board ship when the lights ever go out completely.

The conventional steam propulsion plant is reliable as well as responsive, rugged, and efficient. Operation of the plant is governed by natural laws, and the plant responds to these laws. However, the business of operating a warship is expensive in terms of both fuel and manpower.

Consequently, Navy men get involved with topics and relationships such as superheat and propulsion economy, speed and power, perhaps more than they did in the past. The purpose of this chapter has been in some cases to acquaint you, but in most cases to refresh your memory, with some basic topics of propulsion plant operation. Subsequent chapters will cover topics in greater depth.

Two words of caution. First of all, the primary objective of this book is to discuss with the OOD some of the information he will need to know to make him "more comfortable with engineering." As a result, terms such as "thermal energy" were used to cover many situations, when actually thermal energy consists of both the *internal energy* within a solid, liquid, or vapor, and *flow of energy*, also called heat, which exists when a temperature difference occurs. It should also be noted that large and weighty volumes have been written on the subject of thermal energy, as well as boilers, turbines, and pumps, to name just a few other topics. Lengthy definitions contained in those volumes are not required by the OOD.

The second word of caution is that relatively few people find the subject of propulsion as interesting as, say, topics that are taboo in wardroom conversation. So it may be necessary to review books on the subject of propulsion and engineering periodically. This book will help you, and so will the experts you'll talk to on board ship.

You'll discover that you can be "comfortable" with your propulsion plant.

CHAPTER REFERENCES

Encyclopaedia Britannica, "Thermodynamics"
Introduction to Marine Engineering, Robert F. Latham, United States Naval Institute, 1958
Principles of Naval Engineering, NavPers 10788-B
Shipboard Electrical Systems, NavEdTra 10864-D

2

Plant Insurance Policies

Navy men who are comfortable with engineering say that their propulsion plants are reliable, efficient, and responsive.

A marine propulsion installation with these qualities is also expensive. Some of the original expense of construction goes to purchase "plant insurance policies" to guarantee the plant will remain reliable, efficient, and responsive throughout its lifetime. For the duration of the ship's life—usually in excess of 20 years, and normally about 30 years—the Navy buys additional "policies" and pays premiums on those already in effect to ensure that propulsion dependability is maintained.

Like the basic propulsion plant, plant insurance policies are also expensive. During the warship's existence the costs of these policies and their premiums may exceed the cost of the propulsion plant itself. Plant insurance policies for conventional surface warships inlude the following:

- Equipment duplication, which we considered in the first chapter as a means of ensuring that the basic steam cycle will not be interrupted.
- The Planned Maintenance System (PMS), which has as its goal complete reliability of all equipment and systems in the engineering plant and throughout the ship.
- Automatic controls, used in numerous ships to deliver feedwater to boilers to support all steaming demands and in many ships to regulate the amounts of fuel and air needed for economical combustion.
- Steam system subsidiaries, which ensure that there are adequate amounts of feedwater and auxiliary exhaust for the plant.
- Emergency generators and the emergency electrical distribution system, which act together to prevent the loss of propulsion, steering, and other vital services.

• Proper water quality, which ensures that boilers will generate steam safely and economically.

• Training of engineers and line officers to ensure that each man knows the value of "plant insurance policies," as well as how these policies relate to the plant and to each other.

The relationship between the propulsion plant and any of its insurance policies is summed up by one word: balance. This chapter will be concerned primarily with automatic controls, steam system subsidiaries, the emergency electrical system, and the maintenance of proper water quality, but the other "policies" mentioned above are equally important in maintaining this balance. The propulsion plant operates properly in a balanced condition. Any change upsets this balance temporarily; the propulsion plant insurance policies restore this balance.

When the plant is operating properly—its insurance policies in effect, and their coverage complete—numerous changes go unnoticed to everyone on board ship except the engineers.

As an example, the OOD or junior OOD (JOOD) standing watch can easily detect an increase in the ship's speed from ahead ⅔ to ahead full. Since the automatic boiler controls and the makeup feedwater system are functioning normally, the plant responds without difficulty. More feedwater, air, and fuel are required to support the increased steaming demand, but this is not readily apparent to anyone except engineering watchstanders.

But suppose the insurance provided by the automatic boiler controls (ABCs) and the makeup feed (MUF) systems was not complete, perhaps as a result of inadequate PMS performance. In this case the results could be both visual and dramatic—heavy black smoke in one case, or a boiler low water casualty in another.

In an extreme situation the result of an increase in speed could be visual, dramatic, sudden, and unfortunate. Consider this example:

Water quality has not been maintained. Deposits collect on the watersides of the boiler tubes, impeding the transfer of thermal energy between the firebox and fluid inside the tubes. More fuel is required to heat the water, but the increase in fuel consumption is hardly noticed.

Now the bridge calls for an increase in speed. The ship will be required to make high speeds for a prolonged period. The firing rate increases. The energy (heat) transfer between the firebox and water inside the generating tubes becomes more critical. The excessive waterside deposits slow down this energy transfer, and heat cannot be carried away quickly enough by the water-steam fluid inside the tubes. Metal begins to weaken.

Engineering watch supervisors notice that suddenly much more water is required to maintain the steam drum level. So far, however, the makeup feed system appears capable of supplying the increased demand. The senior boiler technician on watch peers through a protective glass watch port into the firebox—but he fails to notice a pinhole leak in one of the generating tubes about two feet above the brickwork of the furnace floor. Meanwhile the EOOW reports to the OOD that the boiler is using both fuel and water in excessive amounts.

Then suddenly it happens—a massive tube rupture that blows out the fire and sends live steam roaring into the fireroom. Watchstanders dive for the deckplates. They claw their way toward the escape trunk—but some do not escape.

An overdramatization? Possibly, and yet men have died in just such an accident. Still more men have been injured in those rare casualties of this nature. Most casualties are far less serious, but even loss of the electrical load during independent steaming can cost a ship thousands of dollars in failed electronic parts and hundreds of man-hours to replace these parts.

On board ship, the engineers place primary emphasis on preventing casualties by operating their plant as specified in technical manuals, the Engineering Operational Sequencing System (EOSS), and standing orders promulgated by the engineer officer, and in maintaining their plant as required by PMS. The Captain and the engineer officer also allocate certain periods for engineering casualty control training so that the engineers will be able to minimize the effects of a casualty if one does occur.

Plant insurance policies are necessary both to prevent casualties and to minimize the effects of the casualties that do occur. If "coverage" under any of these policies is not extensive enough to protect the plant, there are no beneficiaries. The status of automatic boiler controls, key steam system subsidiaries, the emergency electrical system, and water quality is important to the OOD. He needs to know that "premiums" have been paid to keep his plant insurance policies in effect.

The first of four plant insurance policies to be examined in this chapter are automatic boiler controls (ABCs).

THE ABCs OF ABCs

Today the great majority of the Navy's frigates, destroyers, and cruisers, and nearly half of its aircraft carriers are powered by 1200 PSI propulsion plants. This situation will continue throughout much of the 1980s, even though gas turbine–powered FFGs and DDs will have entered the fleet in large numbers by 1982.

The higher temperatures and pressures and generally smaller water volumes of the 1200 PSI propulsion plant represent a savings in both weight and space over lower pressure plants for a given amount of shaft horsepower. Since it is a conventional steam plant, the 1200 PSI plant is similar in most respects to 450 and 600 PSI propulsion plants. However, the 1200 PSI plant requires such rapid and accurate response to feedwater, fuel, and air demands that automatic controls are necessary for safe operation. (DDs 931–944 have automatic feedwater controls only; all other 1200 PSI warships have automatic feedwater and combustion controls.) Almost all automatic boiler control systems are pneumatic systems, which depend on compressed air and are relatively insensitive to shock.

Here are some advantages of automatic boiler controls—which have pleased the Navy so much that many of the newer conventionally powered amphibious and replenishment ships also are now automated:

1. Men are freed from vital but nevertheless tedious tasks.
2. Man-hours saved during watches can be used for maintenance.
3. Operation is possible under conditions that might temporarily disable fireroom crews, such as the shock wave during an NBC attack or a surface action involving smokescreens.
4. Automatic controls respond to change more rapidly than men and do not overcompensate for change, so that they enhance (fuel) economy.

Disadvantages include more complicated firerooms, dependence on a source of control air, additional maintenance for the control systems, and the possibility of problems caused by calibration errors.

With so much emphasis placed upon the proper operation of automatic controls, it's obvious that the ABC maintenance man and his assistant(s) play an important role in today's propulsion plants. If you asked the ABC maintenance man for some basic information about the control systems in your ship, he would begin by saying that all systems perform the same functions of measuring, comparing, computing, and correcting. Figure 2-1 shows that these functions are performed by transmitters, various types of relays, and controllers.

The next item the ABC maintenance man would explain is that ABCs have the following three modes of operation:

1. *Automatic:* The system maintains steam pressure at a desired value (set point) in either the steam drum or at the superheater outlet in combustion systems; and steam drum level at centerline in feedwater systems. (Centerline is also a set point.)
2. *Remote manual:* The fireroom supervisor takes control of the system at the control panel and introduces air signals into the system to maintain set point(s).
3. *Local manual:* The blowers and fuel to the burners must be controlled by hand. A check watch controls the level of the steam drum. (This is the mode of control for DD 692/710.)

Another common feature of automatic boiler control systems is the fact that almost all systems have *air locks,* which act as commands to the final control elements and hold the steaming rate constant if the system loses its source of control air. In case a particular system either has no air lock or the air lock malfunctions, certain events take place. In *combustion systems* combustion air will go to a minimum value. Since fuel is "slaved" to air to prevent improper combustion and the ship from making black smoke,

Combustion control
Feedwater control
Transmitters
Fuel oil to burners
Fuel oil from burners
Combustion air flow
Steam pressure
Steam flow
Feedwater flow
Steam drum water level
Relays
Fuel oil burned
Steam demand
Feed flow and steam flow
Controllers
Air flow, oil flow, and steam demand
Combustion air flow and steam demand
Steam drum level, feedwater flow, and steam flow
Final control elements
Fuel oil control valve
Forced draft blower steam admission valve
Forced draft blower dampers
Feedwater control valve
Results
Constant steam pressure is maintained with maximum combustion efficiency
Steam drum water level is maintained by correct feedwater flow

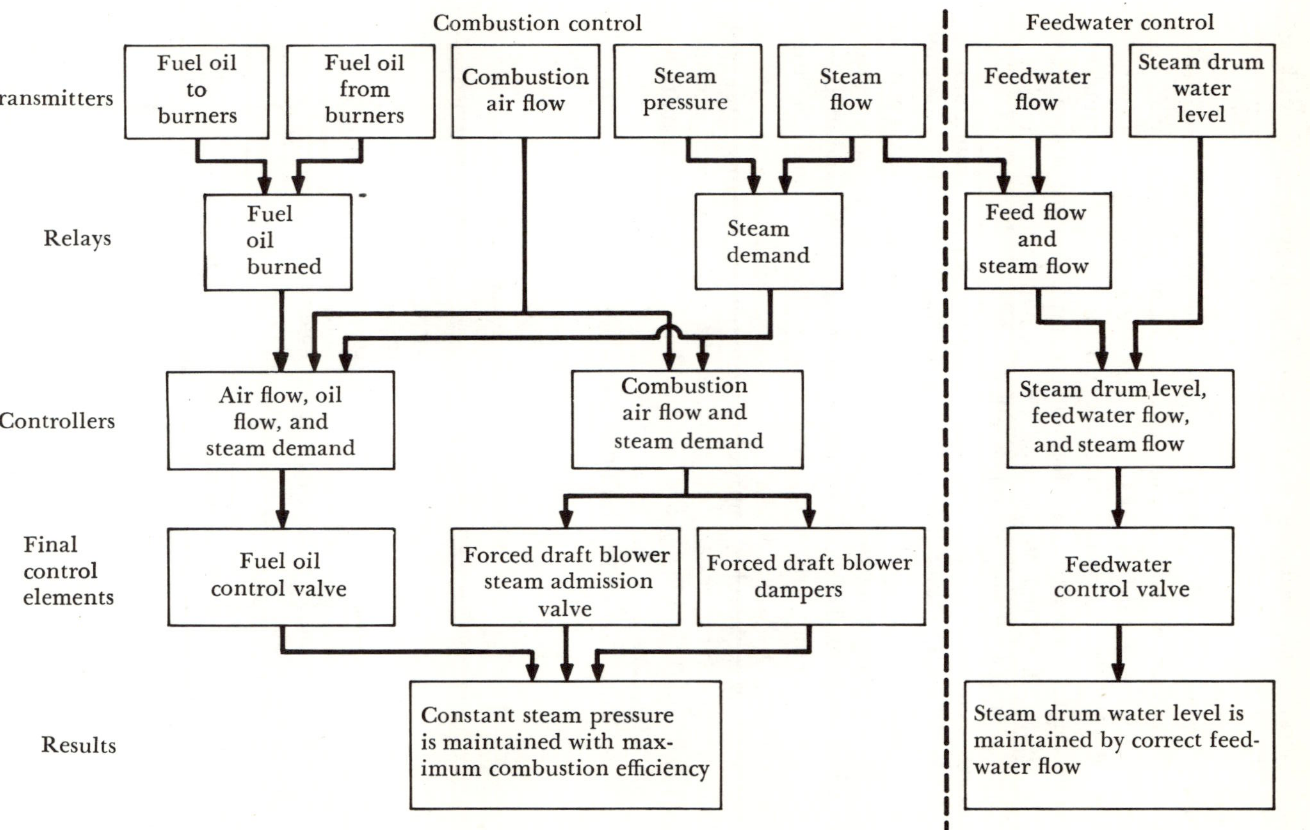

Figure 2–1.

the fuel oil control valve will be secured. In the majority of *feedwater systems,* however, the feedwater control valve opens wide, and quick action by the fireroom watch is necessary to avoid a boiler high water casualty.

Although many consider ABCs to be just one system for a particular ship, ABCs consist of both automatic combustion controls (ACC) and automatic feedwater controls (AFWC). There are various types of each. In the case of 1200 PSI warships, there are four different manufacturers of feedwater controls. However, all feedwater control systems utilize steam flow, feedwater flow, and steam drum level to maintain drum level at centerline to support all steaming demands.

Feedwater controls must compensate for the transient conditions of "shrink" and "swell." Swell occurs during an increase in steam demand (greater steam flow). A greater volume of steam suddenly begins moving through boiler generating tubes, and the water level of the steam drum tends to rise temporarily. Shrink occurs when the steam demand decreases. The volume of steam in the generating tubes is decreased, and the drum level drops temporarily.

If the feedwater control system did not compensate for these transient conditions, a low water casualty would occur during an increased demand, even though "swell" would *tend to indicate* that a high water casualty was imminent. The opposite is true for the "shrink" effect.

Figure 2-2 shows a highly simplified feedwater control system. In this simple—but representative—feedwater control system, the transmitters

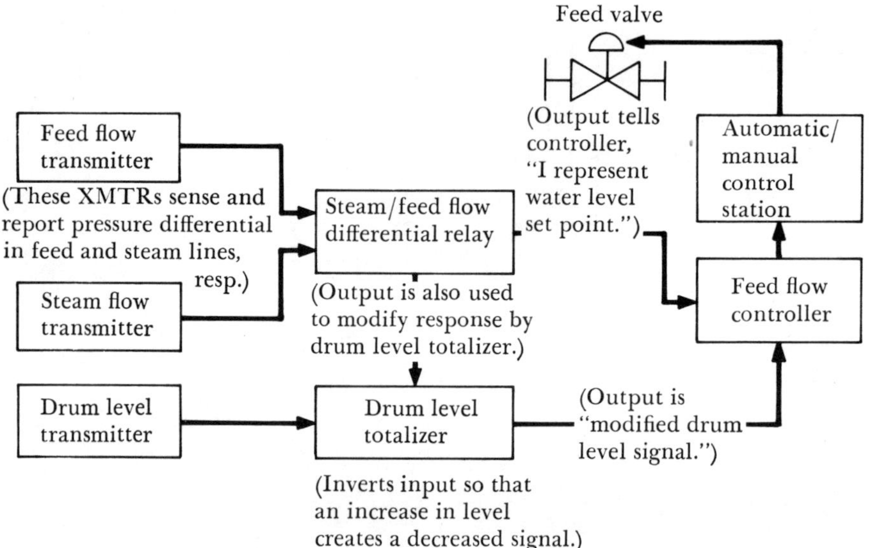

Figure 2–2. Simplified automatic feedwater control

send air at different loading pressures to other components. The drum level totalizer and the steam/feed flow differential relay refine the incoming signals and send their own outputs to the feed flow controller at different pneumatic loading pressures. The feed flow controller accepts both signals and sends its own pneumatic output signal through the automatic/manual (A/M) control station to a positioner on the feedwater flow control valve, which responds to the order.

If the system is in *automatic*, this loading signal will pass through the A/M control station. If the system appears to be malfunctioning, however, the watch supervisor will position a knob at the A/M control station and go to *remote manual* operation. This allows him to block the controller's signal and to introduce his own signal to the positioner on the feedwater flow control valve.

But perhaps it's easier to look at the situation the way a young officer does when the chief BT explains the ABCs of ABCs on their duty night: "Think of the feed flow controller as the feedwater check valve man. He wants to turn the valve wheel when his eyes notice a change in steam drum level, just like the steam/feed flow differential relay says to do. But experience tells him to wait for shrink or swell to subside—just like the totalizer tells the controller."

The chief can also compare the simplified combustion control ABC (Figure 2-3) to the jobs his men on watch would have to do in *local manual* operation. Here's the way the chief explains it:

"The steam pressure controller is really the top watch, the most important man. When the bridge rings up a bell (speed change) the information from that boiler pressure gage tells him what to do.

"Say the bell calls for an increase in steam demand. If the top watch let the burnerman cut in more burners right away, the ship would smoke black. Instead the supervisor immediately orders the blowerman—in this system he's named the air flow controller—to increase the speed of the forced air draft blowers. But the burnerman has to wait.

"The inertia of those forced draft blowers takes them a little while to respond. But soon the increased windbox pressure tells the watch that there's sufficient air now to prevent smoking black. Then the top watch orders the burnerman to cut in more burners; you might say that the top watch just acted as the low pressure selector would in an ABC system.

"When the steam pressure returns to normal, or set point, the blowerman watches his periscope and makes minor adjustments for good combustion, the way an air flow/oil flow ratio relay would do. The burnerman is helping maintain steam pressure with minor micrometer valve adjustments so the oil flow and oil pressure maintain a workable relationship."

Then the chief summarizes: "That fireroom crew has worked together for some time now. They're good, too. But they aren't as quick and as accurate as their ABCs, and they tire a great deal faster."

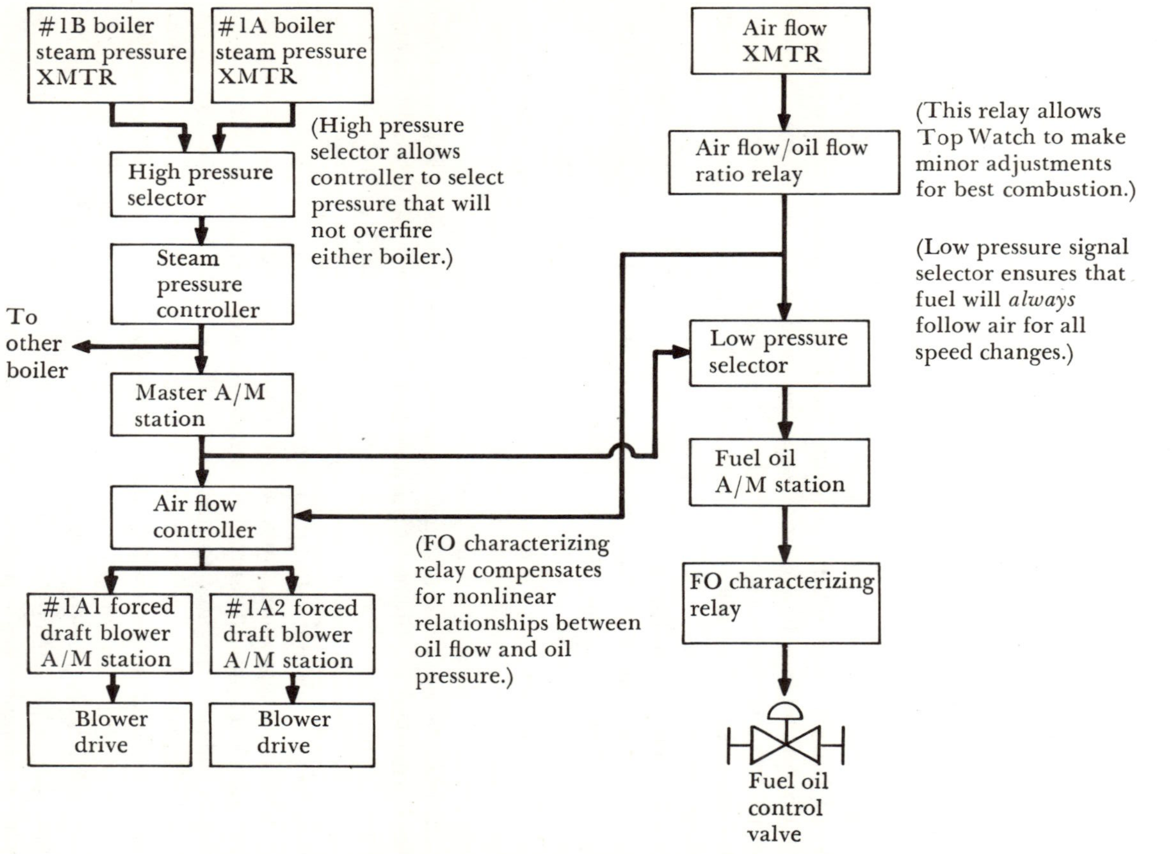

Figure 2–3. Simplified automatic combustion control

Shipboard engineering, like most other subjects, tends to have its own language. ABCs don't clarify this situation much, since there are three major manufacturers, and variations in even their products. The master sender in a Hagan ABC system shares many things in common with a controller in the General Regulator system, but it may be used more like a Bailey Meter Standatrol in other ways. Technical discussions can be confusing.

ABCs, regardless of their design, enhance plant responsiveness and reliability, efficiency, and economy. ABCs are such an important plant insurance policy that in the case of 1200 PSI warships in particular, OODs are cautioned against attempting rapid and radical speed changes unless feedwater control is in full automatic.

Still, only so much reliance can be placed on any one insurance policy, even so important a policy as ABCs. NavShips *Technical Manual,* Chapter 221, points out the following: "Casualty conditions may necessitate transfer to manual (Feedwater) check valve operation. To ensure competence in this type of operation, a sufficient time at least once each week should be taken to train and drill each watch in transfer and manual check valve operation."

Commanding officers and their chief engineers put well-trained watch sections high on their lists of necessary plant insurance policies. Steam system subsidiaries are also on these lists.

STEAM SYSTEM SUBSIDIARIES

Although conventional steam propulsion plants are "closed cycle" plants, some fluid losses inevitably occur from the time fluid leaves the boiler as steam until fluid returns to the boiler as feedwater. All plants require a readily available source of feedwater to make up for these losses. Naval propulsion plants also require a constant source of auxiliary exhaust, which provides the thermal energy necessary to distill feedwater and potable water from seawater. (Auxiliary exhaust also "scrubs" condensate free of entrapped air and serves as a heating agent as condensate becomes feedwater in the DFT.)

Both the makeup and excess feed system and the auxiliary exhaust system are important to the OOD because of the relationship of each to the main steam system.

In the majority of 1200 PSI ships the makeup and excess feed system operates automatically. A drop in the water level of the DFT is sensed by the system, causing an air pilot–operated valve to open. This permits water from the makeup feed tank (also called the reserve feed tank) on suction to enter either the main condenser or the fresh water drain collecting tank, depending on ship class and the lineup of the system. Although the amount of makeup feedwater required depends upon the DFT level, no feedwater enters the DFT directly from the reserve feed (makeup feed) tank. Instead, makeup is taken on through the condensate system via the main condenser or fresh water drain collecting tank.

Periodically, the DFT may contain too much water, since one of its functions is to act as a surge tank in the basic steam cycle. This could cause incomplete deaeration. When the water level in the DFT becomes excessive, an air pilot–operated excess feed valve diverts some condensate to the same reserve feed tank from which makeup feed is taken.

Some of the earlier 1200 PSI plants and many of lower pressure propulsion plants do not have this type of levelmatic (automatic) control. In these classes of ships—and in the newer ships during casualty situations—the watch must control the level of the DFT manually.

Bridge watch officers need to know relatively little about the "nuts and bolts" operation of the makeup and excess feed system, but they do have a passing interest in its performance. Each day the Fuel and Water Report tells them how much reserve feedwater was introduced into the steam cycle to make up for losses during the previous 24 hours. This amount varies daily. OODs who make a habit of checking this daily report before it goes to the Captain know that a consistently high water usage indicates that the propulsion plant is not "tight." High usage also means that more fuel must be burned to heat the relatively cool reserve feedwater that is introduced into the steam cycle to compensate for system losses.

The auxiliary exhaust system, which rivals the makeup and excess feed system in importance, is the second steam system subsidiary of concern to the OOD.

Auxiliary exhaust is used to heat and deaerate condensate before it can be delivered to the boiler as feedwater. The distilling units (evaporators) also require auxiliary exhuast as the heat source for converting seawater to water for men and propulsion on conventionally powered ships. As a result, auxiliary exhaust is so important that its constant supply must be guaranteed. It's almost a case of a system insurance policy requiring insurance policies of its own.

Augmenting stations ensure that auxiliary exhaust will always be available under normal conditions. Augmenters similar to those found on board a 1200 PSI DDG class ship, listed in Chapter 1, would be set up for operation as indicated in the following typical example:

The desired pressure of the auxiliary exhaust system is 15 psi. If pressure drops below desired pressure, two pressure-reducing stations augmented from the 150 PSI auxiliary system will operate. One reducing valve in the fireroom would open at 14 psi, while another reducing valve in the engineroom would begin to operate at 15 psi. If auxiliary exhaust line pressure continued to drop, a pressure-reducing valve in the fireroom 1200 PSI DSH line would begin to operate at 12 psi.

Excessive pressures in the auxiliary exhaust line must also be avoided—that DFT shell looks big, but the DFTs plating is not very thick. Moreover, the DFT must operate as designed to ensure feedwater deaeration. On board 1200 PSI ships an increase in pressure will cause an air pilot–operated valve to open and excessive auxiliary exhaust to dump into the main condenser. This valve, in the engineroom, normally is set to open at about 18 psi and to close at 16 psi. If the auxiliary exhaust line pressure is still too great, relief valves in both the engineroom and the fireroom will direct the excessive exhaust to the atmosphere via escape piping in the stacks. (That's the wisp of steam sometimes seen coming from the stacks of ships inport.) While it is not necessary to become concerned about exact pressures in the auxiliary exhaust system, it does help to have an appreciation of the protection afforded the propulsion plant by both the auxiliary exhaust system and the makeup and excess feed system.

Now it's time for a look at another plant insurance policy, a policy so important that the Navy would not own a ship without one—the emergency electrical generation and distribution system.

EMERGENCY ELECTRICAL GENERATION AND DISTRIBUTION SYSTEM

Nearly every book written about shipboard engineering contains this or a similar statement: "A ship without electrical power has no capability as a fighting unit."

Even if the ship is not a warship, an emergency electrical generation and distribution system (emergency power system) is needed to regain propulsion and to retain vital services. Vital services for any ship include pumps which will support the propulsion plant, emergency lighting, steering, and the means to tell friendly units of the ship's problem. In the case of a warship, vital services also includes the ability to point at least one weapon at an opponent during an engagement—if the warship has any hope of living and fighting again another day.

On board ship the emergency electrical generation and distribution system, also called the emergency power system (EPS), and the casualty power system (CPS) provide back-up insurance for the ship service (SS) system. Line officers should be more familiar with the emergency power system, since this system is the primary back-up for the ship service system.

The emergency power system consists of at least one emergency generator, an associated emergency switchboard, and the emergency feeders, which transfer the emergency power to selected vital loads in the event of a casualty to the ship service system.

In naval ships physical separation and isolation are used to ensure that the emergency power system is not affected by a casualty which results in loss of ship service power. For example, on board a destroyer the *normal* and *alternate* ship service feeders are located below the waterline on opposite sides of the ship. Physical separation of these large ship service feeders is just one device employed by ship designers to enable a ship to withstand battle damage. The *emergency* feeder is located near the centerline and well above the normal waterline in these ships to protect further against loss of vital services.

The emergency generator and its associated switchboard (SWBD) are also isolated from the ship service generators and their associated switchboards. This is true even in the case of the 1200 PSI FF/FFGs, which have a diesel-powered generator (SSDG) that can be used as either a ship service generator or an emergency generator. (In these class ships the diesel-powered generator is routinely used as a ship service generator inport.)

In some ships, such as aircraft carriers, the size of the ship requires a system of vertical zone control for SS and emergency distribution. In these ships a number of vertical zones are established, with each containing a *load center* SWBD which is energized by a *bus feeder* from the ship service switchboard. The load center switchboard supplies power to the electrical loads in its zone.

Emergency switchboards may supply more than one of these SS zones, since the number of emergency zones depends on the number of emergency generators installed. The USS *Enterprise* (CVN 65), for example, has four large emergency diesel generators.

Diesel or gas turbine engines are used as the prime movers of emergency generators on board conventionally powered surface ships.

The rated output of an emergency generator varies according to ship class, but in general the capacity of the emergency power system on board DD 931 and DDG-2 class ships is more critical than in other installations. The reason for this is the relatively low output of these generators (100 KW, 160 amps) compared to the reliance that must be placed on electrical auxiliaries to regain propulsion in either of these ships in a casualty involving the complete loss of steam.

The simplified illustration of Figure 2-4 represents the interconnection between the ship service and the emergency power systems in DDs, DDGs and 1200 PSI CGs.

The illustration shows that both emergency switchboards can be energized from either SWBD. (The nearer source is designated the *preferred* or *normal* source in most installations.) In case neither the *normal/preferred* nor *alternate* sources can energize the emergency switchboard, the associated emergency generator starts automatically, and the emergency SWBD is transferred to it.

This transfer is accomplished by an automatic bus transfer (ABT). In some ships the emergency SWBD will transfer back to its preferred or

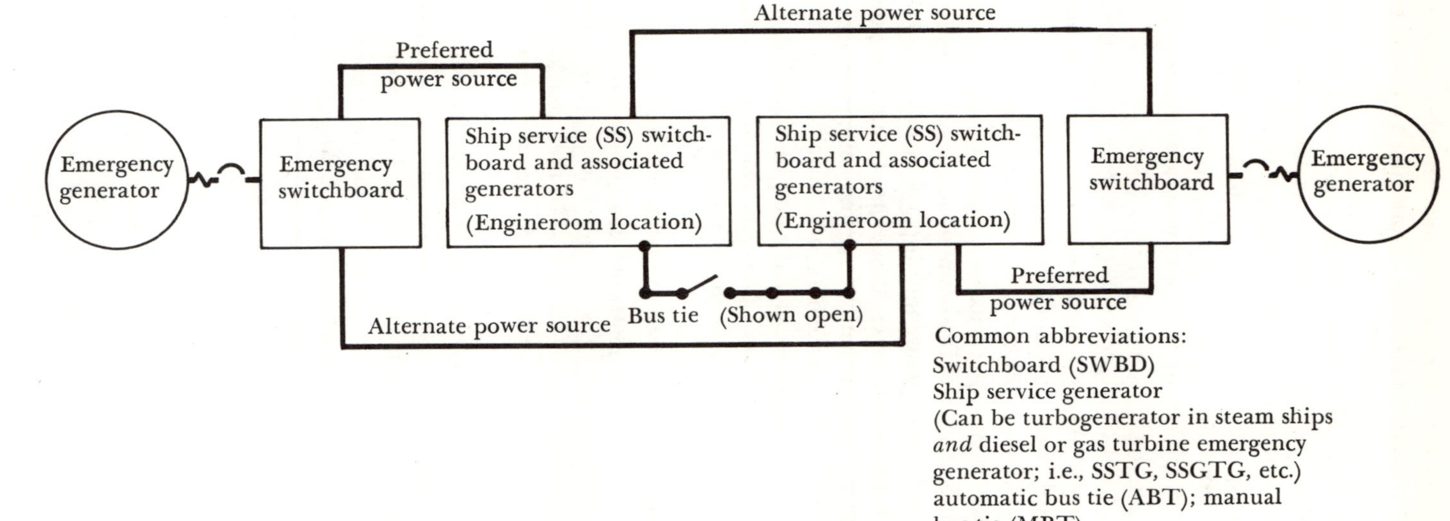

Common abbreviations:
Switchboard (SWBD)
Ship service generator
(Can be turbogenerator in steam ships *and* diesel or gas turbine emergency generator; i.e., SSTG, SSGTG, etc.)
automatic bus tie (ABT); manual bus tie (MBT)

Figure 2–4. Simplified 1200 PSI DD/DDG power distribution

alternate source if either becomes available. In other ships the emergency generator must be tripped manually before the ABT will transfer to the preferred or alternate source.

Here's an example of how the emergency power system works:

A twin-screw ship is under way, with firerooms supplying steam to the main engine and a turbogenerator only in their associated plant (split plant operation). The electrical load is also split, since the bus ties between the forward and aft ship service switchboards are open.

A boiler casualty occurs in the forward plant. Before the electrician's mate on watch could close his switchboard bus ties, the forward emergency switchboard sensed a problem. (Emergency switchboards are to be energized under way.) As soon as voltage available from the forward ship service switchboard dropped below 300 volts, an ABT on the forward emergency switchboard transferred the forward emergency switchboard's power source to the after ship service switchboard.

Now a second boiler casualty happens—this time in the after plant. Vital electrical loads throughout the ship begin to trip off the line. The after emergency switchboard losses its preferred source of power; it has no alternate source of power as a result of the first boiler casualty.

At this point ABTs at both emergency switchboards activate the *automatic start* features of their associated emergency generators. This occurs within 10 seconds, and emergency power is available to vital equipments to regain propulsion and safety of the ship. Figure 2-5 represents the final steps in this chain of events.

From the discussion above, bridge watch officers will want to be sure that the EOOW knows the emergency generators are lined up for automatic start at all times under way, and that air banks are charged for instantaneous starting of these generators. In the event of a complete loss of the normal ship service generators, a *feedback tie* from the emergency switchboard to a ship service switchboard will enable an emergency generator to supply power to selected portions of the ship service switchboard load; this, too, is a worthy topic for discussion between a prospective OOD and the EOOW and senior electrician's mate on board ship.

The emergency power system protects vital equipments necessary to support the basic steam cycle and ship safety. This valuable plant insurance policy will be required if the premiums on still another plant insurance policy—proper water quality—have not been maintained.

WATER QUALITY

The high operating temperatures and pressures of naval boilers require that water quality be maintained within exact limits in order to prevent deterioration of boiler metals. Boiler water treatment is a great deal more scientific today than it was years ago, when engineers put potato peelings

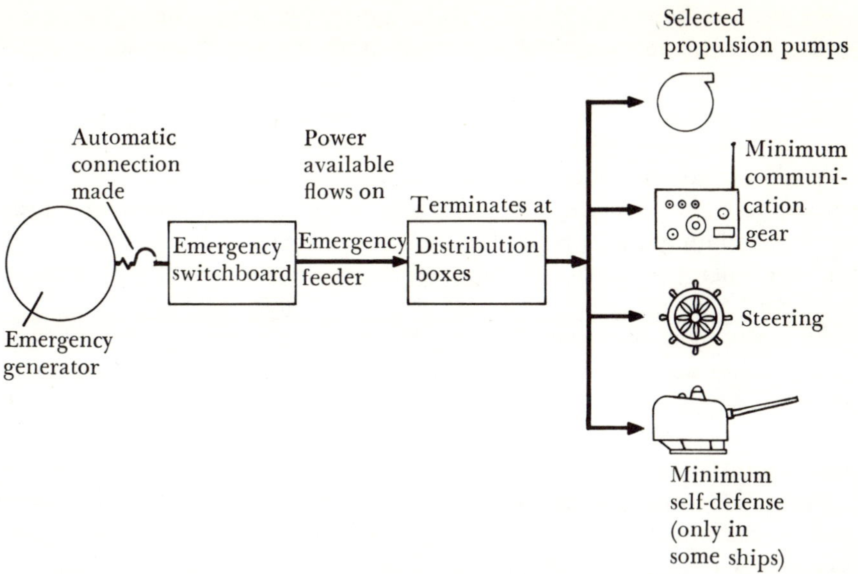

Figure 2–5. Emergency power for selected vital loads

into the water inside Scotch fire tube boilers to prevent excessive accumulations of scale on the watersides of the tubes.

Today few subjects in the engineering department draw more attention than the maintenance of proper boiler water quality. The oil king (water king in larger ships) is the engineer officer's principal assistant in this task. Few men on board ship perform a job as important as his.

According to naval training manuals, the transformation of seawater into distillate by the ship's distilling units (also called the "evaps" by the engineers) still results in water that contains 70 pounds of sea salts for every 20,000 tons of distillate. That does not seem like much. However, this water, or distillate, entraps some oxygen whenever the two come together. Distillate also contains minute quantities of metal from the tanks in which it is stored.

Dissolved oxygen and sea salt contamination would make boiler water treatment necessary even if no other contaminants fouled the water. However, seawater can also enter the water cycle through small leaks in heat exchangers such as the main condenser and through funnel drains, drain collecting tanks, lines that run through the bilges, and tiny leaks that could exist in makeup (reserve) feed tanks. Lube oil can enter the water cycle through the turbine gland seal system. Even boiler water treatment chemicals add to the total amount of foreign material present in the water cycle.

Figure 2-6 shows that the greatest concentration of contaminants occurs in the boiler. These contaminants produce three major categories of problems that must be controlled: waterside deposits, waterside corrosion, and carryover.

Waterside deposits interfere with heat transfer. Unless boiler waterside deposits are controlled, larger amounts of fuel will be required to produce steam, and plant efficiency suffers. High firing rates can eventually produce tube failures, putting boilers out of commission and threatening the safety of men on watch. (In this case the fireside surface of the metal is weakened.)

Waterside deposits include insoluble deposits, resulting in sludges, which tend to collect in the bottom of the boiler; calcium and magnesium salts, forming scale, which is difficult to remove but, fortunately, rare; and corrosion deposits, which are the result of chemical reactions between metal and water.

Waterside corrosion can cause tube failures which begin with tube metal weakening from the inside. This occurs, according to present theories, as a result of the electrolytic action, which can be controlled but never completely eliminated in any boiler.

Waterside corrosion is caused by localized pitting, which occurs as a result of dissolved oxygen in the water, and general corrosion, which results in a loss of metal over an entire surface.

General corrosion occurs when boiler water is too alkaline (the pH value too high in 1200 PSI boilers), when alkalinity is too low (low pH), or if the chloride content of the boiler water is too high.

Carryover may actually be the first indication that proper water quality is not being maintained. (Carryover may also be caused by the fireroom watch supervisor raising the steam drum level too high before conducting a surface blowdown or because a maintenance crew assembled the internal fittings in the steam drum incorrectly. However, this section refers to carryover as a result of failure to maintain water quality.)

Carryover occurs whenever the amount of dissolved or suspended solids in the steam drum is so great that the bubbles of steam have difficulty breaking free from the steam/water mixture. Water is then carried along with the steam into the superheater and possibly beyond that. *Priming* is the carryover of large amounts of water. Priming is particularly dangerous because this "wet" steam has a relatively lower temperature which can cause thermal shock to turbines.

Fortunately, a conscientious oil king and his effective program can maintain boiler water quality. He collects samples and tests water from the boiler(s), feed tanks, and DFT as often as required by the NavShips *Technical Manual* chapter on water treatment. After testing, the oil king conducts his treatments with *blowdowns* and with *chemical treatment.*

Blowdowns involve opening valves to remove water from the boiler. A

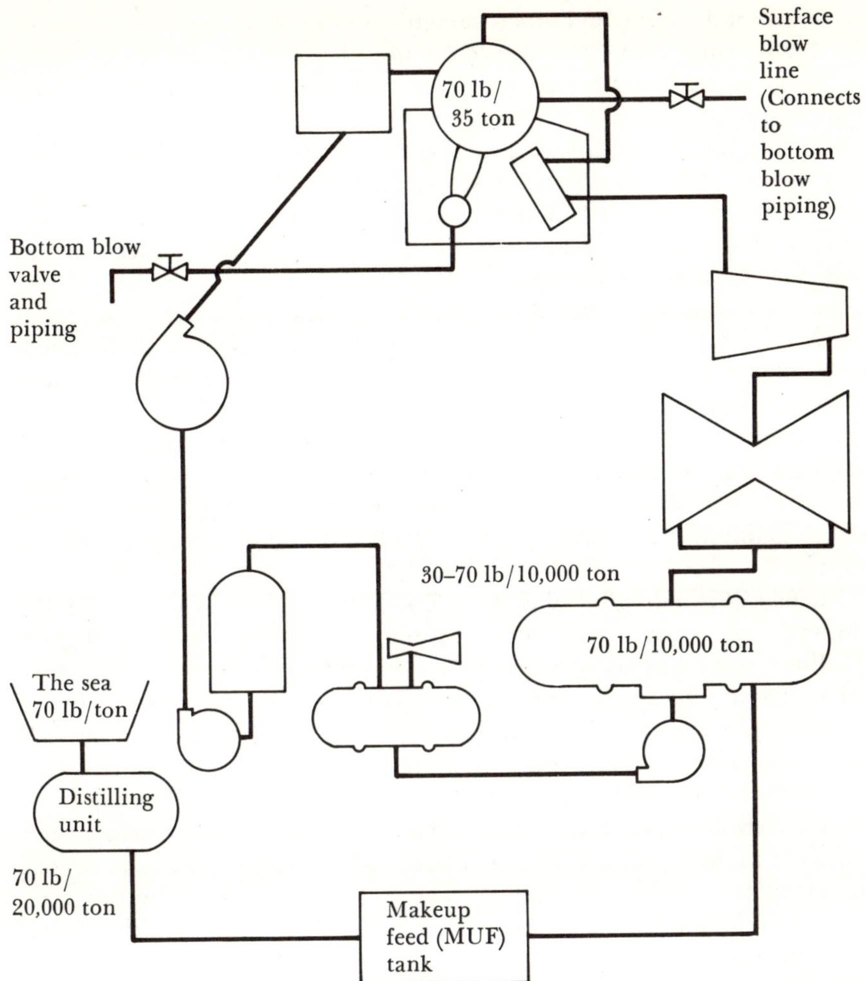

(MUF is "pulled" into the main condenser by vacuum drag in some 1200 PSI plants and pumped into the main condenser from the fresh water drain collecting tank in other 1200 PSI plants, depending on ship class.)

Figure 2–6. Contaminants in the water cycle

blowdown represents an unavoidable loss of water, but blowdowns are necessary to remove contaminants. During the blowdown the oil king and his assistants open and then secure blowdown valves and the overboard (skin) valve in a specified sequence in order to prevent injury to personnel and to conduct a thorough blowdown.

Daily, and more often if required, the oil king conducts a 10% *surface blowdown*. This is done with the boiler on the line to remove chlorides, dissolved solids, scum, and other foreign material which tends to collect in the steam drum. *The OOD should beware of making speed changes once he has given the EOOW permission for a surface blow of the boiler feeding the affected shaft.*

The oil king also gives each steaming boiler a *bottom blowdown* at least once weekly. The boiler must be secured for *bottom blow* and allowed to cool for at least one hour. Bottom blows are used to control the accumulations of sludge in the mud (water) drum and waterside headers and to decrease total boiler contamination.

In twin-screw ships the usual practice during peacetime is to *cross-connect* the plant and steam with one boiler feeding both shafts while conducting bottom blowdowns. Each boiler should also receive a bottom blow upon being secured.

To conduct chemical treatment, the oil king tests water and then adds chemicals in the amounts required to completely eliminate hardness (a property water acquires because of the presence of certain dissolved salts) and to maintain the alkalinity (pH in 1200 PSI boilers) within prescribed limits.

For 600 PSI boilers he adds Navy Boiler Compound to counteract the impurities derived from sea salts, and to give the boiler water the slightly alkaline range required by NavShips *Technical Manual.*

For 1200 PSI boilers he adds caustic soda to place the boiler water in the positive pH range required to resist general acid corrosion. Then he adds the proper amount of disodium phosphate so that any scale-forming salts remaining will be converted to sludge.

ODDs should review the Daily Fuel and Water Report before they submit it to the Captain or command duty officer with the other 12 o'clock reports. These reports are self-explanatory. The report lists the following normal limits for boilers required by NavShips to ensure effective boiler water treatment:

600 PSI Boilers			*1200 PSI Boilers*
2.0 equivalents/million	Chloride		2.0 equivalents/million
2.5–3.5 epm	Alkalinity	pH	10.4–11.0
0	Hardness	Phosphate	10–25 parts/million
1300 micromhos	Total Conductivity (Measure of total contamination)		700 micromhos

Some engineer officers require that the oil king indicate that the results of the tests he records show if the trend in each category is up or down.

The oil king also supervises *lay up* of idle boilers. Failure to lay up idle boilers properly allows dissolved oxygen to attack boiler metals; over a period of time the results of improper lay up can be as disastrous to a boiler as feedwater that received improper deaeration in the deaerating feed tank. The corrosive effects of dissolved oxygen can become catastrophic in high-temperature, high-pressure boilers. The choice of a dry lay up or a wet lay up depends upon a number of factors. Under way, idle boilers are almost always maintained in a wet lay up, with water level in the steam drum at normal level, steam in the remaining portion of the steam drum, and lay up of the superheater as specified by the boiler technical manual.

In the following chapter we will see why proper water quality plays such an important part in overall fireroom operation.

Still, proper water quality is only one of the important propulsion plant insurance policies. In this chapter we have also examined basic automatic controls, key steam systems subsidiaries, and the emergency electrical generation and distribution system. None is more important than proper training of engineers or proper maintenance procedures according to PMS, subjects which are beyond the scope of this book.

Understanding what the key plant insurance policies are, however, and knowing that "payment" of the premiums on each is current, can help a prospective OOD be comfortable with engineering.

CHAPTER REFERENCES

Engineering Operation and Maintenance, NavPers 10813-B
Principles of Naval Engineering, NavPers 10788-B
Shipboard Electrical Systems, NavEdTra 10864-D

3

The Fireroom—Where It's Made

Some chief engineers refer to their boilers as the "bread and butter" of the propulsion plant.

This concept highlights the importance of the boiler and its associated fireroom equipment. In a very real sense those engineer officers are correct—the process that eventually results in the conversion of thermal energy to mechanical energy and then into work begins in the boiler. To place fireroom operation in its proper perspective, reconsider the economic analogy used in Chapter 1:

The boiler is a manufacturing plant—producing steam for sale—to buyers consisting of the main engines, turbogenerators, and various pumps.

Figure 3-1, a diagram dealing with the functional relationships of propulsion units, auxiliaries, and the various steam systems, illustrates the importance of the boiler and associated fireroom equipment. This figure shows that a boiler casualty or any casualty that interferes with the generation process can have a serious effect upon propulsion. Such a casualty could also affect the ship's electrical plant, at least temporarily.

When the engineering plant operates as designed, equipments and procedures in the fireroom cause the OOD and JOOD little concern.

But the combination of an equipment malfunction—possibly because of lack of effective PMS—and inexperienced watch personnel changes the situation rapidly. Two of the most feared boiler casualties, high water and low water, can develop in less than 60 seconds at high steaming rates. Still another casualty, a ruptured boiler tube, rarely gives advance warning.

Moreover, one casualty often triggers another. Some engineers refer to

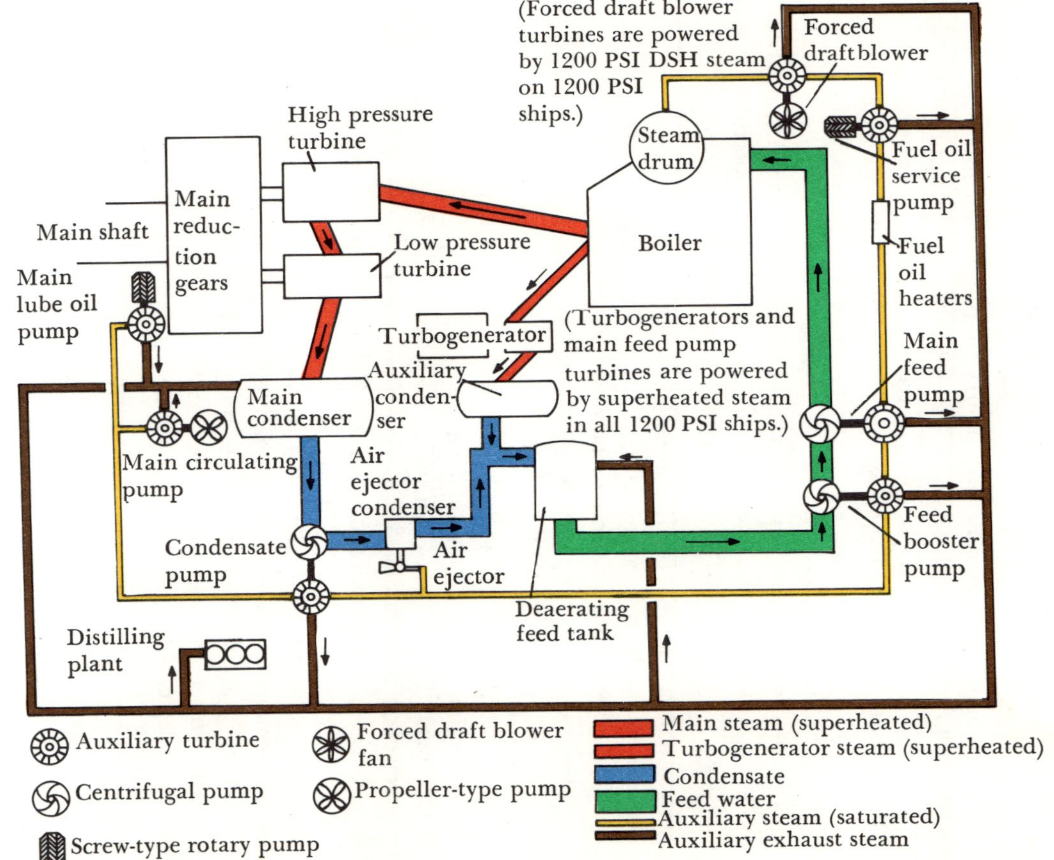

Figure 3–1. Functional relationships of propulsion units, auxiliaries, main steam system, and auxiliary exhaust system

this as the "domino effect," a highly descriptive phrase. Engineering casualty control consists of breaking the chain of casualties and regaining control of the plant as quickly as possible.

A deck watch officer, the OOD or the JOOD, must be much more than an interested bystander during a fireroom casualty or any other engineering casualty. He needs to understand how an engineering casualty can occur so he can help limit its effects and enable the engineers to regain control of his plant.

As an example, a knowledgeable OOD isn't one who disregards the EOOW's report that the fireroom crew is having difficulty maintaining steam pressure, and orders a speed increase instead. A knowledgeable OOD knows that a speed increase must be supported by more steam for the turbines of various pumps and the forced draft blowers, which will put the propulsion plant even further in jeopardy.

Officers of the deck, and the men who aspire to become OODs, need to understand important fireroom relationships in order to perform with confidence. This chapter will cover the following topics:

Feed system operational relationships
Feed system–boiler related casualties
Fuel and air systems operational relationships
Fuel and air systems–boiler related casualties
ABC–boiler related casualties

A final section will categorize fireroom casualties and recommend courses of action for OODs on single and twin-plant ships.

Before beginning with an outline of feed system operational relationships, however, it is important to note that the U. S. Navy has many different classes of conventionally powered surface ships. As a result, boilers, fuel oil service pumps, and forced draft blowers (superchargers for combustion air in FF-1040 and FFG-1 class ships) are the only equipments common to many firerooms. Equipment configuration varies. For example, the fireroom of an AOR contains three boilers; the fireroom of a destroyer, frigate, or cruiser contains two boilers; and a main machinery room of a conventionally powered aircraft carrier may contain two boilers as well as the associated main engine and its shafting.

There are numerous differences even in destroyer firerooms. On board 600 PSI destroyers the deaerating feed tank (DFT), main feed booster pumps, and main feed pumps are located not in the firerooms, but rather in the enginerooms. But the relatively smaller water volumes of 1200 PSI boilers require rapid reaction to feed system problems, so the DFT, main feed booster pumps, and main feed pumps are located in the firerooms of 1200 PSI DD/DDGs, FF/FFGs and the 1200 PSI cruisers.

Even the equipment in a 1200 PSI fireroom varies. Most DD-931 class ships have only two main feed booster pumps and two main feed pumps in either fireroom. All other 1200 PSI DDs, DDGs, and CGs have three main feed booster pumps and three main feed pumps in both firerooms. Yet there can be differences even here, since some ship classes have three motor-driven main feed booster pumps, while another ship with an identical plant layout may have one turbine-driven and two motor-driven main feed booster pumps.

Fuel oil service pumps (FOSP) in 1200 PSI firerooms vary according to ship class. The FOSPs in FF firerooms are motor-driven, while 1200 PSI DD/DDG/CGs have two turbine-driven main fuel oil service pumps and a smaller electric pump for inport use in each plant.

The important point to remember here is that regardless of equipment configuration, your engineering plant has a great degree of reliability. Sufficient duplication of equipment exists throughout the engineering plant to ensure that any warship can fulfill her mobility missions and support her combat mission areas. Propulsion plant insurance policies, including well-trained engineers and a properly functioning PMS program, enhance this reliability.

A knowledgeable OOD enhances the reliability of his propulsion plant also. Most of the examples used in this and the following chapter concern 1200 PSI plants, but the principles of steam propulsion are the same for any ship. You can be a knowledgeable OOD who is "comfortable" with engineering.

FEED SYSTEM OPERATIONAL RELATIONSHIPS

Conventional steam propulsion plants presently power two-thirds of the U. S. Navy's surface fleet. The great majority of these conventional plants for years to come will be 1200 PSI propulsion plants.

1200 PSI plants share many similarities with the men who operate these plants. They are capable of great reliability and responsiveness, but unforgiving if ignored. Men and propulsion plants are also sensitive to change. The 1200 PSI plant in particular cannot tolerate an interruption in its feedwater supply.

To prevent this, designers moved the DFT, the feed booster pumps, and the main feed pumps into the firerooms of 1200 PSI ships. This gives the boiler technicians (BTs) control of the feed system, although they still must share responsibility for control of the condensate system with the machinist's mates (MMs). (This sometimes leads to problems caused in part by lack of communications; but, more of that later.)

Regardless of who "owns" the feed pumps, BTs and MMs in any ship agree that the feed pumps are vital. A senior BT conducting a tour of his fireroom can describe the problems that could lead to the loss of a feed pump—overspeeding, loss of lube oil pressure, close tolerances leading to the possibility of overheating, and failure of the booster pump to provide suction.

Various safety devices protect the main feed pump if any of these conditions exist. However, this "protection" usually results in securing of the pump. This creates the possibility of the "domino effect." Here's a typical situation:

The ship is making 15 knots during a normal transit, with one boiler supplying steam to its associated main engine. One main feed pump is on the line, the second is turning over in a standby condition, and the third has been placed out of commission for PMS. The machinist's mates have one condensate pump on the line in the engineroom. Soon the bridge will order a major speed change; the OOD does not inform the EOOW.

The plant responds to the ordered speed. However, the level of the DFT begins to drop—hard steaming over the past two months has resulted in the deferral of major maintenance on the condensate pump presently in use; now the maximum discharge pressure of this condensate pump has been greatly reduced. The makeup feed system, a plant insurance policy, attempts to return DFT level to normal. Still the DFT water level continues to drop.

The BT top watch calls the EOOW to report he cannot maintain desired level in the DFT. At this point the EOOW belatedly orders the engineroom to put its second condensate pump on the line. But in his haste to comply, the engineroom lower level watch opens the discharge valve on this second condensate pump as wide as possible.

The sudden increase of relatively cold condensate destroys the critical balance between temperature and pressure existing in the DFT. DFT shell pressure drops immediately. The main feed booster pump requires a positive suction pressure from the DFT. When the main feed booster pump "senses" this loss of suction pressure, the feed water fluid in the booster pump suddenly "flashes" to steam.

Now the low suction safety trip feature of the main feed pump senses a greatly decreased discharge pressure from the booster pump. If the booster pump cannot supply sufficient pressure, feedwater in the main feed pump will also "flash" into steam. Design of the main feed pump requires that feedwater, rather than steam, be used to dissipate the large amount of heat created by the feed pump's rapid rotation as it increases the feed pressure necessary to force feedwater into the steaming boiler. The main feed pump now faces the danger of "seizing," and possibly burning up. So the low suction safety trip activates to secure the main feed pump.

And a boiler low water casualty—one of the most potentially devastating of all engineering casualties—results.

Engineers of a multi-plant ship reply that they can always *cross-connect* major steam systems to maintain mobility and services, the primary goal of casualty control. (When the engineers cross-connect, they operate designated valves in a prescribed sequence to allow main engines, turbogenerators, and other steam "users" to receive steam from another in-service boiler. Numerous other systems on board a warship can be cross-connected, such as the firemain, and electrical power, which is done by electrician's mates operating transfer devices at distribution switchboards. Cross-connecting is normally a casualty control procedure that permits a prime mover or other equipment to receive energy necessary for its operation; the term will be used again in this and later chapters.)

But in a single-plant ship, such as an FF or FFG-1, a boiler low water casualty (assuming only one of the two propulsion boilers is in operation) results in loss of propulsion. Additionally, only operation of the ship service diesel generator (SSDG) remains before emergency battery-powered lighting/total darkness.

Fortunately, proper use of certain procedures provides a type of "insurance" against the possibility of a low water casualty. These procedures include proper plant lineup, use of the reserve feed transfer pump in multi-plant 1200 PSI warships, and—in an extreme emergency—use of "cold suction."

Approved PMS and Engineering Operational Sequencing System (EOSS) procedures for main feed pump lineup require that low suction safety trips have different settings; so a decreasing booster pressure will not mean the loss of all main feed pumps at once. BT's follow these procedures when lining up their fireroom.

BTs in multi-plant 1200 PSI ships may also use the *reserve feed transfer pump* to transfer water directly from a feed tank to the DFT. (Unlike 600 PSI DDs, 1200 PSI DD/DDG/CGs cannot cross-connect the main feed system between the forward and after propulsion plants.)

BT's in all 1200 PSI warships may also attempt to feed the boiler with *cold suction* in an extreme emergency. This procedure requires that a motor-driven feed booster pump be lined up directly to a reserve feed tank. If the booster pumps in normal lineup with the DFT lose suction, the booster pump lined up for cold suction starts automatically. However, use of either the reserve feed transfer pump to pump up the DFT or cold suction is a temporary measure. Unless the watch can stabilize plant conditions, a low water casualty may soon result.

It should be noted that many senior engineers dislike lining up their plants for the emergency use of cold suction, even though this procedure has been approved by the Naval Sea Systems Command.

They contend that cold water from a reserve feed tank will damage their main feed pump because of thermal shock—"jump right off its foundation!" is a phrase often used when these men describe the possible effects of cold suction on the main feed pump.

They further contend that even if the main feed pump remains operational, the cold, undeaerated water from a reserve feed tank will almost certainly damage their boilers.

The key point to remember is that use of cold suction is an *emergency* condition. Thermal shock which accompanies the use of cold suction could damage a main feed pump and possibly the boiler. However, safety of a warship alongside another during underway replenishment or maneuvering radically to avoid an incoming cruise missile is more important than the possible loss of a main feed pump.

OODs and their EOOWs should be able to speak to each other as professionals who understand what constitutes both normal and emergency conditions. The primary objective of engineering casualty control is to maintain mobility and services. Accomplishment of PMS, proper training, and understanding when cold suction might be necessary are only some of the ingredients necessary to maintain propulsion and services and to limit damage in engineering casualty control.

Most commanding officers do not require that their OODs memorize the names of protective safety features, such as the low suction safety trip of the main feed pump, for equipments in the feed system. Exact names such as speed limiting governor of the main feed pump (speed limiting governors are also used to limit the speed of the turbines of numerous other pumps in the engineering plant) or the DFT vacuum breaker are *not* too important to the OOD—but he should realize that safety features do exist to protect equipments in the feed system.

The OOD should also realize that malfunction of *any* equipment in the feed system could have disastrous results for steam generation. Proper operation of the DFT and feed pumps is so important that in 1200 PSI warships observation and control of the DFT and feed pumps is the primary responsibility of the fireroom *upper level/pump watch.* (Other watchstanders perform the same function in other classes of ships.)

If the ABC feedwater control in warships with ABCs becomes inoperative, a *checkman* must be stationed. His *only* responsibility is maintaining proper boiler water level.

FEED SYSTEM–BOILER RELATED CASUALTIES

Low water, high water, and a *ruptured tube* are the three most common feed system–related boiler casualties.

Unless the engineers act quickly, loss of either the DFT, feed suction, or a main feed pump results in the high probability of a low water casualty. In ships without ABCs, lack of attention on the part of the checkman can also cause a low water.

A high water casualty can occur from priming, or when ABC control air is lost. If a ship has ABCs and the feedwater control loop is inoperative, a checkman must be stationed. The occurrence of a high water casualty has the same probability as a low water casualty when the checkman becomes inattentive.

In addition, controlling the entry of feed water by manually opening/closing the feedwater check valve can be a difficult task in any 1200 PSI ship, even for experienced petty officers.

A ruptured tube rarely gives advance warning. Steaming a boiler with its water "out of limits" for prolonged periods creates the possibility of a ruptured tube occurring soon.

All three casualties threaten the immediate safety of men and equipment.

As used on board ship, the term "low water" refers to progressive conditions of a drop in steam drum water level. The first condition occurs when the water level, as sighted in the boiler gauge glass, drops to a point which activates the *low water level alarm.* (The level at which this alarm actuates varies among ship classes.) The second condition occurs when the water level in the boiler gauge glass drops completely out of sight. At this point the boiler is considered to have suffered a low water casualty, and the possibility that the boiler's design steam generation has been adversely affected cannot be discounted.

When the alarm condition occurs, the boiler technicians must react quickly to secure the boiler to prevent the second condition, the actual low water casualty.

A low water casualty can be extremely damaging to the boiler. The furnace is hot, and there is an insufficient amount of water circulating through the boiler to carry away the heat. Unless the boiler is secured, the boiler's tube surfaces, brickwork, and air casing will be seriously damaged.

The most dangerous action the fireroom watch could take in a low water casualty would be to continue to feed the boiler. Dangerous steam or water leaks could occur because of the sudden temperature imbalance.

A low water casualty requires the engineer officer and the chief BT to inspect the boiler after it cools. Even if the boiler has not been damaged, it will still be out of commission for 48–72 hours.

Similarly, the term "high water" refers to both an alarm condition and a casualty.

Engineroom watchstanders who have experienced a serious high water casualty report that a high water sounds like a freight train roaring through the main steam lines.

The intensity of the sound only hints at the potential violence of a high water casualty. The engineroom watch must quickly secure steam to the main engine and turbogenerator(s) before blading is ripped from the turbine wheels and slashes through the turbine casings.

Priming is a form of high water caused by failure to maintain water quality, which we discussed in Chapter 2.

Carryover, a problem that may be associated with water quality, is a form of high water which has cumulative, rather than dramatic, effects. This condition results in steamside deposits in the superheater, in steam lines, and on turbine blading. Carryover is characterized by low superheat temperatures in the engineroom and a general loss in plant efficiency.

The effects of a ruptured tube casualty vary in intensity. In some cases a pinhole leak can be detected only because the DFT continually demands a large quantity of makeup feedwater, or the BTs detect a change in the flame pattern in the boiler firebox.

In more extreme cases a ruptured tube results in an immediate low water, and steam blows out the fire and roars into the fireroom.

Like high water and low water, even the pinhole variety of ruptured tube is serious. The fireroom watch will secure the boiler immediately.

Effective casualty control depends upon many factors. The ability of the *entire watch team* to recognize abnormal conditions is one of these factors.

For example, the upper level watch in a 1200 PSI DD/DDG/FF/FFG/CG fireroom must monitor the operation of the DFT, main feed pumps, and combustion air equipment as well as periodically observe the steam drum water level.

Suppose, in this example, that either a high water or low water condition developed and the respective alarm did not sound. By the time he noted the lack of a water level indication in the boiler gauge glass, the upper level watch would not be able to tell whether the drum level was too high or too low. At this point he would shout: "Water out of sight!" The BT top watch, also known as the fireroom supervisor, would then order the boiler secured.

In this example, an abnormal condition could also be noticed first by either the burnerman or the fireroom messenger. Perhaps the burnerman might notice a sudden fluctuation in steam pressure. Perhaps the messenger might suddenly find he could not ascertain the level in the remote (steam drum) water level indicator. The man who first noted an abnormal condition would quickly and loudly inform the watch supervisor using standard phraseology, and immediate casualty control procedures would begin.

In one case, the first indication the engineering gang in the cruiser on which I served as engineer officer received of a pinhole tube rupture came when the OOD called the EOOW to report a strange noise coming from the forward stack.

The OOD reported the noise sound rather like the whistle of a far-off freight train. From wardroom training sessions the OOD, a warrant officer whose billet was the ship's tactical data systems maintenance officer, recalled that such a noise can be symptomatic of a pinhole tube rupture. So the OOD called to compare notes with the EOOW. . . .

The ship was fortunate that this OOD, who usually dealt in electrons, knew enough about his engineering plant to become "uncomfortable" when he sensed an abnormal condition.

FUEL AND AIR SYSTEMS OPERATIONAL RELATIONSHIPS

Fuel System

The first energy transformation in a conventional propulsion plant occurs with the conversion of chemical energy to thermal energy during the process of combustion.

Obviously, the supply of both air and fuel to the boiler in proper amounts of each is important to steam generation. Equipment duplication ensures the availability of fuel and air for combustion. There are at least two fuel oil service pumps (FOSP) in every propulsion plant, and at least two forced draft blowers (FDB) are available to supply combustion air to each boiler. (The only exception to the amount of equipment devoted to air supply occurs in the FF-1040 and FFG-1 class warships. These warships have a single source of air for each 1200 PSI pressure-fired steam generator, an air compressor that supplies combustion air under pressure and is driven by a gas turbine [not to be confused with the typical propulsion gas turbine described in Chapter 8]. The air compressor and the gas turbine together are referred to as the *supercharger*.)

The basic boiler fuel oil system on board ship includes fuel oil storage tanks, fuel oil service tanks, fuel oil piping, fuel oil service pumps, fuel oil strainers, various valves, and the various types of burner configurations. Figure 3-2 shows a very basic fuel oil service system of interest to the OOD.

Before the fuel oil service system illustrated in Figure 3-2 is explained, it should be noted that the OOD must realize the importance of the fuel oil storage tanks and the contaminated oil settling tank (one in each fireroom), which are not shown. Both the storage tanks and the contaminated tank are considered parts of the fuel oil system.

The OOD is concerned with fuel oil storage tanks because stability of the ship decreases if these tanks are allowed to remain empty as fuel is transferred from them to service tanks. (In some of the newer class ships, such as the FF-1040, FFG-1 and DD-963 classes, storage tanks are never empty, since seawater ballast automatically fills them as fuel is used. These ballast systems are called seawater compensating systems.) In calm seas this decrease in stability poses no great problem. In heavy weather, however, a warship with empty fuel oil storage tanks surrenders a potentially vital degree of stability.

(An explanation of ballast systems is not given in this book, since a ballast system as such has no part in contributing to the propulsion of a

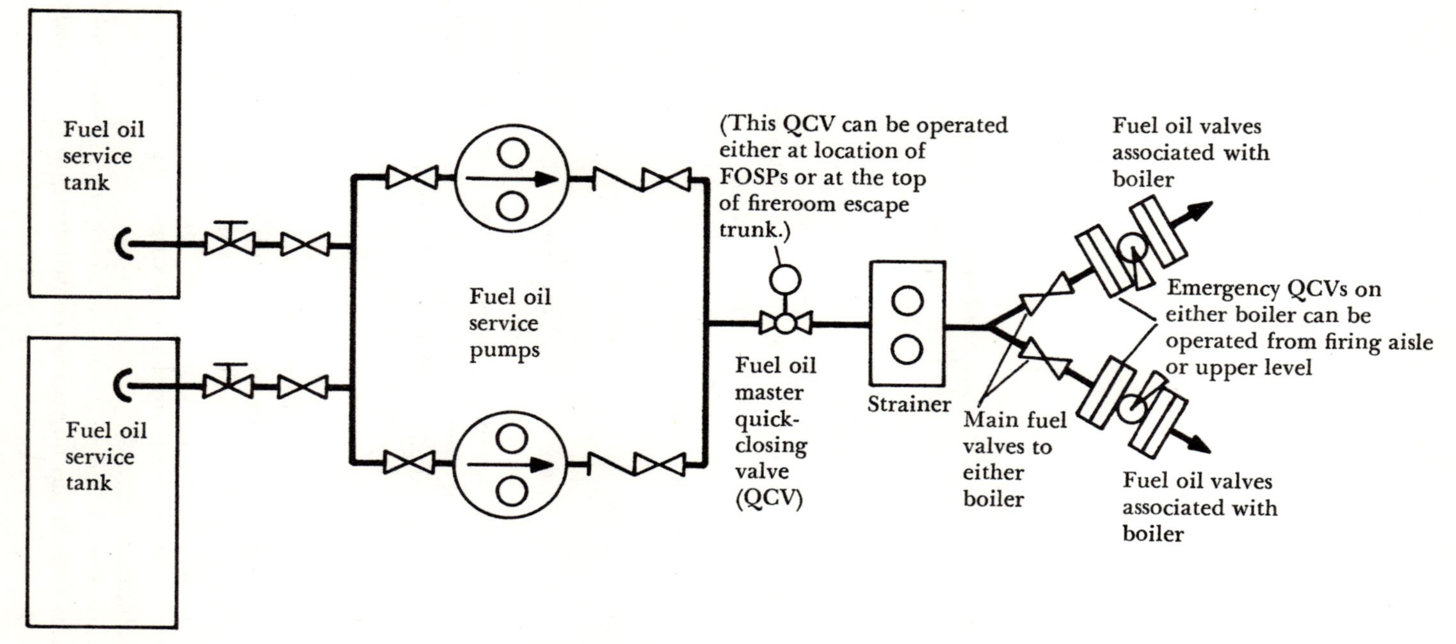

Figure 3–2. Simplified fuel oil service system

ship. However, the prospective OOD should realize that time is required to ballast, that even more time is required to deballast prior to receiving fuel, and that whenever a ship takes on seawater ballast, the possibility of water contamination of the fuel oil occurs.)

The contaminated oil settling tank is required in casualties involving water contamination of fuel oil. Water in the fuel oil is a serious casualty, which requires diverting the contaminated oil to the contaminated oil settling tank until a sample shows the fuel oil to be free of water or a different line up of the fuel oil service system can be effected.

As shown in Figure 3-2, there are at least two fuel oil service pumps for each propulsion plant of a warship. In most ships each FOSP may be lined up with its suction isolated on either fuel oil service tank. In general, one FOSP is sufficient to supply all steaming demands required of a propulsion plant, even when both boilers are on the line. In such a case, however, good engineering practice would require the second pump to be ready in a standby condition.

Good engineering practice also requires that both fuel oil service tanks in each plant remain at least 50% full at all times. The oil king "tops off" service tanks at least daily—and much more often at high steaming rates—by transferring fuel oil from storage tanks to service tanks with the fuel oil transfer pump. Fuel oil suction is shifted whenever a service tank drops below 50% of its capacity. Before shifting suction, the oil king uses the fuel oil stripping system to clear the service tank being readied for use of any water that may have collected in it. Then he takes a "thief sample" to ensure that no water is present in the fuel oil service tank, which he will now place on suction.

The FOSPs for use at sea on board most naval ships are turbine-driven. However, the increasing reliance on motor-driven auxiliaries on all the 1200 PSI FFs and FFGs has resulted in the fuel oil service pumps of these ships also being electrical rather than steam-driven.

In order to ensure continuation of fuel oil service, in case the electrical load is lost temporarily, nitrogen-charged accumulators are installed in the fuel oil service systems of all 1200 PSI FF/FFGs. If the FOSP trips off the line because of loss of the electrical load or pump failure, the accumulator will "push" sufficient fuel into the furnace to maintain fires for a brief period. During this interval—normally about 20 seconds—the ship service diesel/emergency (SSDG) generator should be up to speed and electrical power available to vital propulsion plant auxiliary equipment.

After being discharged from the fuel oil service pump, fuel oil passes through a master quick-closing valve (QCV) on its way to the strainer. This QCV provides an important casualty control feature that allows fuel to be cut off either at the pump location or at the top of the fireroom escape trunk in case a catastrophic casualty requires the fireroom to be evacuated. (The fact that fuel oil to a steaming boiler can be secured from this remote location is also important to the OOD inport, as will be explained in a later chapter.)

The fuel oil strainer is a simple piece of equipment in the fuel oil service system, but an important one. The strainer will free the oil of any particles that could clog the burner equipment, permitting improper combustion and robbing the boiler of its efficiency. The BT top watch will order strainer baskets to be shifted if the pressure drop across the strainer exceeds 3% of the pump discharge pressure.

From the strainer, fuel oil service piping directs the fuel oil to either or both of the boilers in the fireroom. Fuel to the boiler passes through a main fuel valve and then through the emergency boiler quick-closing valve on its way to valves at the boiler front. The boiler emergency QCV is still another casualty control feature; this valve can be actuated either at the boiler front from the "firing alley" or from the upper level.

After passing through the boiler emergency QCV, fuel oil enters a configuration of fuel oil valves corresponding to the type of combustion system employed. Fuel oil is mixed with air supplied by the forced draft blower, or air compressor/turbocharger in the case of pressure-fired steam generators, in the burner assembly, and combustion results.

Air System

A proper balance between fuel and air is critical to plant efficiency and even safety. Generally speaking, it takes about 15.5 pounds of air to furnish the oxygen required for the combustion of one pound of fuel.

According to the Navy's fuel experts, approximately 12% of the total energy available in a given quantity of fossil fuel is wasted when carbon monoxide, rather than carbon dioxide, is a by-product of combustion. To say it more simply, you're getting inefficient combustion—and that can affect your ability to continue surveillance of a foreign cruiser group or to arrive safely at the next port before a storm.

Fortunately, the BTs can obtain proper combustion by providing 10–15% excess air at various firing rates. That excess air saves that 12% of energy that might otherwise be wasted in each barrel of fuel—which, incidentally, costs the country a great deal more than it used to.

On the other hand, too much excess air serves no useful purpose. Instead, too much excess air absorbs heat and carries thermal energy up the stack. Moreover, the forced draft blowers are wasting steam by furnishing more air than necessary for proper combustion.

Propulsion boilers on board the majority of warships of the U. S. Navy require two forced draft blowers for each boiler to supply the massive quantities of air necessary to support high steaming rates. The only propulsion boiler with a single source of air is the 1200 PSI pressure-fired steam generator. Two of these steam generators are installed in the propulsion plant of each FF-1040 and FFG-1 class ship (and the USS *Glover*, an ASW research ship).

The source of combustion air for each of these steam generators is its associated air compressor–gas turbine supercharger. These pressure-fired steam generating installations require only half the space and weigh half as much as conventional boilers with equal generating capabilities.

(A number of Soviet and French warships also have pressure-fired steam generators, according to current editions of *Jane's Fighting Ships*.)

Vertically mounted forced draft blowers are installed in 1200 PSI cruisers, guided missile destroyers, and destroyers. The construction of these forced draft blowers, driven by 1200 psi desuperheated steam like the forced draft blowers of all 1200 PSI warships, creates two special problems:

The first concerns steam leakage along the vertical shaft with the possibility of water condensing in lube oil sumps and on bearings. The BTs must observe strict precautions when lighting off and operating these forced draft blowers to prevent wiping a bearing because of contaminated lube oil.

The second problem can be a result of the first: These vertically mounted forced draft blowers are virtually impossible to repair under way, even in calm waters at slow speeds, because of the combination of heavy machinery, vertical mounting, and finite clearances.

It is unnecessary for the OOD or prospective OOD to become deeply involved in the mechanics of how a forced draft blower operates. But the OOD and JOOD do need to know some specific facts about forced draft blowers. The most important consideration is how forced draft blower availability can affect the maximum speed of their ship.

As an example, a 931-class destroyer and most 1200 PSI DDGs and CGs can make about 27 knots by steaming with one boiler and both of its forced draft blowers on the line in each fireroom. The ship will be steaming *split plant* in such a case.

If any one forced draft blower must be taken out of commission, the maximum speed available drops to about 24 knots.

If the ship steams cross-connected with one boiler (with both forced draft blowers) feeding each main engine, the maximum speed available drops to about 21 knots.

If the ship steams cross-connected on just one boiler and one of its associated forced draft blowers fails, maximum speed available drops to about 17 knots. (However, such a situation would result in some weighty consultation between the Captain and the engineer officer, and the ship would probably be making no more than about 12 knots with preparations under way to light off another boiler!)

The subjects of forced draft blower availability and boiler steaming demands are related to the subject of fuel usage discussed previously. In 1200 PSI destroyers and many other twin-plant warships, the use of two boilers to make a sustained speed of 20 knots is more economical in terms of fuel usage than steaming just one boiler near its maximum firing rate in cross-connected operation to make the same speed. However, it may be more economical in terms of manpower to steam one boiler safely near its maximum limit to perform needed maintenance in the idle fireroom.

Before the relationships of fuel and air systems to boiler casualties are discussed, here are some additional points the OOD should remember concerning forced draft blowers and combustion air supply:

- Men or ABC combustion controls regulate the air and oil mixture and seek to provide efficient combustion at all firing rates.
- Fuel supply should always lag air supply for efficient combustion. (The engineers' phrase "Fuel follows air" may be the most concise form of this statement!)
- In ships with ABC combustion control (and feedwater control), men can always override the automatic controls.
- Heavy black smoke belching from the stack does more than raise the "boiling point" of the Captain, the executive officer, or the boatswain's mates as they admire their newly painted decks. Something could be seriously wrong in the fireroom, and a casualty *could* be imminent.
- A small amount of black smoke means that inefficient combustion is occurring. Stack monitoring devices—which consist of a BT peering up a smoke periscope in most ships—may require checking.
- White smoke is rarely seen, since nearly 200% excess air is required before white smoke is produced. But even under clear stack conditions the forced draft blowers could be running faster than required—pushing too much air into the firebox, raising superheat temperatures higher than required in some types of boilers, and wasting thermal energy up the stack.

FUEL AND AIR SYSTEMS–BOILER RELATED CASUALTIES

Either of two basic unsafe boiler conditions results from a serious imbalance or an interruption in the amounts of air and fuel supplied for combustion. One unsafe condition would be a casualty characterized by a *rapidly dropping steam pressure.* The other unsafe condition creates the possibility of a *flareback.* In either situation, the fireroom watch must take quick action to secure the boiler.

(The term "rapidly dropping steam pressure" is arbitrarily defined, by the author, as an *abnormal* situation with potentially serious consequences and recognized as such by the EOOW or watch supervisor. It is difficult to quantify this term, so put yourself in the place of the EOOW who must recognize a rapidly dropping steam pressure situation:

As an example, the EOOW knows that he can expect steam pressure to drop momentarily from, say, 1200 psi to 1185 psi as his ship increases speed from 25 knots to 27 knots. He also has extreme confidence in his automatic boiler controls (ABCs), which he knows will return steam pressure back to its desired value of 1200 psi within 10–15 seconds. Then, five minutes later and with the ship at 27 knots, steam pressure to a main engine begins to drop . . . 1190 (psi), then 1160, then 1135.

Perhaps only 25 seconds have elapsed since the EOOW ordered the throttleman to "conserve steam" and requested the fireroom top watch to report the status of the fireroom. Perhaps another 15 seconds will pass before the fireroom reports, during which time pressure may now have decreased to 1100 psi; and regardless of whether he has been peering at various gauges, his wristwatch, or the anxious looks on the faces of others, the EOOW knows that *he* is faced with a casualty involving a rapidly dropping steam pressure.)

A casualty characterized by a rapidly dropping steam pressure can result from the following casualties: loss of combustion air, loss of fuel oil suction, fuel oil service pump failure, and water in the fuel oil.

A flareback may occur whenever the pressure in the boiler furnace (firebox) exceeds the air pressure in the boiler air casing. This could be the result of an interruption in the amount of combustion air supply, a delay in lighting the mixture of air and oil in the firebox, or other reasons. Here are some situations likely to result in a flareback:

1. Attempting to relight burners from hot brickwork, rather than using a torch.

2. Forced draft blower failure.

3. Failure to *purge* the furnace after a casualty or during casualty control drills, which results in the accumulation of unburned fuel oil or combustible gases in the furnace, tube bank, economizer, uptake, or air casing.

4. During hostile action, the impact or even near miss of an enemy's bomb or cruise missile, which could create a partial vacuum at the forced draft blower intake. (In such action the fireroom GQ team would make every attempt to keep their boilers operational, but they would also stay as clear as practical from the possible path of a flareback.)

5. An abnormal condition such as water in the fuel oil, which can cause loss of fires. However, some fuel is still getting into the furnace, resulting in its vaporization and a potential explosion.

The fact that water in the fuel oil can result in either a flareback or a dropping steam pressure—and possibly both—makes water in the fuel oil a particularly interesting casualty:

How water in the fuel oil can result in a flareback was described in a preceding paragraph. But suppose the water content was not sufficient to extinguish the fire. Instead, steam pressure begins to drop. The ABCs sense the dropping steam pressure. Now the forced draft blower speeds up, and the (turbine-driven) fuel oil service pump tries to deliver more oil.

Now the steam demand on the boiler is even greater. The boiler's ability to supply this increased steam demand has, of course, already been greatly diminished as a result of the water-contaminated fuel oil. Unless the fireroom watch can secure their boiler when they detect water in the fuel oil, a catastrophic boiler casualty may result.

When the EOOW reports the securing of a boiler as a result of rapidly dropping steam pressure caused, in this case, by water in the fuel oil, the OOD can get an unpleasant feeling of isolation. (This can occur during any casualty!) But the OOD also knows certain events in the engineering plant are taking place:

As soon as the engineroom throttle watch noted steam pressure dropping, he informed the engineroom watch supervisor. That supervisor directed the throttleman to "Standby your throttle to *conserve steam.*" Simultaneously, the EOOW received the report in standard phraseology that the fireroom had suffered a casualty involving water in the fuel oil. The EOOW knew that the fireroom watch's *mandatory* action in securing the boiler would conserve steam in the fireroom, enabling the fireroom watch to get their boiler back on the line as quickly as possible after shifting to the standby fuel oil service tank and obtaining uncontaminated fuel oil.

AUTOMATIC BOILER CONTROLS (ABC)–RELATED BOILER CASUALTIES

ABCs are designed to provide efficient and safe boiler operation, and there are relatively few problems with their use that concern the OOD.

The BT top watch or console operator can override the automatic controls by switching to *remote manual* operation. *Local manual* operation can also be performed safely with proper coordination.

Most of the problems associated with ABCs that can affect the safe operation of boilers result from air failures. Most ABC systems incorporate air locks, which lock the last loading signal received into the control element at the time of the control air failure.

In the event of a control air failure, the OOD should avoid making speed changes or allowing other departments to light off major equipment such as gun mounts without first informing the EOOW. Failure of the OOD and EOOW to exchange information during a loss of ABC control air can easily result in a boiler casualty.

STEAM CATAPULT OPERATION/SUPERHEAT LIGHT OFF ON M-TYPE BOILERS/SAFETY VALVES/BLOWING TUBES

Although information contained in this chapter will benefit the OOD or prospective OOD of a 1200 PSI warship more than others, an objective has been to keep the information factual but general enough to benefit others. Officers who aspire to become OODs in any ship may find information on the following topics of some interest.

Steam Catapult Operation

Engineer officers of aircraft carriers report that there are special problems associated with steam accumulators during catapult launchings.

These problems occur because of the sudden and large steam demands made by the catapults while the accumulators are being charged. These steam demands are similar to those made on a boiler feeding a propulsion plant during radical maneuvering.

ABCs adapt well to boiler installations supporting catapult launchings because of the wide ranges in steam demand. Water level is extremely important because emergency steam demands may result in excessive carryover for extended periods.

Loss of a boiler devoted to catapult launchings usually results in a lower speed available to the carrier. This occurs since aircraft carrier engineer officers normally want to isolate a boiler for accumulator charging, instead of cross-connecting from boilers being used for propulsion.

Superheater Light Off on Controlled Superheat Boilers

The M-type double furnace boiler on board most 600 PSI destroyers and other ships presents a special problem concerning superheater light off. At high firing rates the pressure in the air casing and the furnace cavity make it dangerous to light off burners on the superheat side.

The BTs require advance notice to light off superheat burners safely. Otherwise, the boiler's firing rate must be reduced.

Safety Valves

A propulsion boiler may not be steamed unless all of its safety valves operate in sequence at settings specified by the Naval Sea Systems Command.

The only other normal safety valve concerns for the OOD or JOOD come when the valves are being set or else actuated to relieve pressures that may build up during radical maneuvering. At such times all hands should remain well clear of the stacks.

Upon occasion watch officers may observe a steady plume of steam less than three feet long in calm air coming from the stack's atmospheric exhaust or escape piping. This may in fact be leakage from a safety valve in the maximum amount authorized by NavSea (*NSTM*, Chapter 221,

Paragraph 3.108). However, that plume of steam represents a loss of valuable feedwater.

Blowing Tubes

Soot blowers are installed on each propulsion boiler, except pressure-fired steam generators, to remove depositis of soot from the firesides of a steaming boiler. Soot blowers must be used regularly and in proper sequence to prevent the accumulation of carbon on generating, superheater, and economizer tubes. Excessive deposits of soot will interfere with the transfer of thermal energy to the fluid inside the tubes. In the case of the economizer, excessive deposits could also result in a fire hazard.

The process of using the soot blowers is called "blowing tubes." Since blowing tubes could result in covering the decks with soot or increasing the chances of a ship being located during periods of electronic silence, the engineer officer of the watch must obtain permission from the OOD before blowing tubes. (Tubes should also be blown weekly in port as prescribed by the senior officer present afloat [SOPA].)

When permission is obtained from the OOD (who may have to alter course to minimize the soot falling upon the weatherdecks), soot blowers are operated in sequence. Steam at a reduced pressure from each soot blower sweeps the soot progressively toward the boiler uptakes. The fireroom supervisor will increase the speed of his forced draft blowers slightly to make certain all soot deposits are removed from the boiler while blowing tubes.

FIREROOM CASUALTIES AND THE OOD

The OOD should epitomize confidence. But inport or under way, *any* OOD can develop feelings of isolation when an engineering casualty occurs.

Fireroom casualties can be categorized. If a watch officer understands a problem exists, he knows what that problem may cause. This causal relationship *may* eventually end with a mandatory action, such as the securing of a boiler. But even then the OOD knows what to expect and what to do—with confidence.

His role in the total casualty control effort can prevent too many dominoes from tumbling. There are, however, some points to remember:

• The goal of the engineers in a casualty situation is to act safely to restore services and to regain propulsion as soon as possible.

• The terms "high water, low water, and dropping steam pressure" describe *either* a condition that causes (an) alarm or else a point requiring mandatory action. For example, water that drops below the bottom of a steam drum gauge glass requires mandatory securing of the boiler. Before the water dropped that low, however, a low water level alarm activated, while the watch was attempting to return water level to normal.

(The key point is that the EOOW's warning of engineering problems requires prudent action by the OOD, but not necessarily an emergency action if plant conditions can be stabilized.)

- Engineers attempting to restore services and propulsion follow set procedures and have minimum time for conversation.
- Poor communication procedures between the bridge and main control deteriorate even more during crisis situations.

Table 3-1 shows the main fireroom type casualties that can affect any conventional steam propulsion plant.

Recommended OOD Actions at Sea—Fireroom Casualties

Commanding officers of all naval ships require that their line officers understand fully the Captain's desires for ship control safety before qualifying as the OOD. This section, therefore, presents only recommended courses of action by the OOD during a fireroom casualty.

A similar section on engineroom casualties is presented in Chapter 4.

The response time of the engineers in controlling casualties depends on a number of factors. These include the existence of air-operated main and auxiliary steam stop valves in some ships, the degree of automation in the plant, physical layout of engineering spaces, the effectiveness of intraship communications, and—most important—the training and expertise of the engineers themselves.

OOD—Single-Plant Ship

(This assumes that the ship has been steaming only one boiler or a casualty occurs which affects all propulsion boilers.)

If the casualty involves a dropping steam pressure, slow to the maximum *safe tactical speed* immediately.

If boiler(s) must be secured, pass the word to safeguard all electronic equipment, turn rudder to avoid ships, and position the ship the best possible way with regard to wind and sea. Inform the OTC using the most secure method possible.

OOD—Twin-Plant Ship

(This assumes the ship has been steaming only one boiler in each plant, or that a casualty which occurs in one plant affects both boilers.)

If the casualty involves a dropping steam pressure, slow to minimum safe speed on the affected shaft.

If boiler(s) in one plant must be secured, slow to *emergency standard*

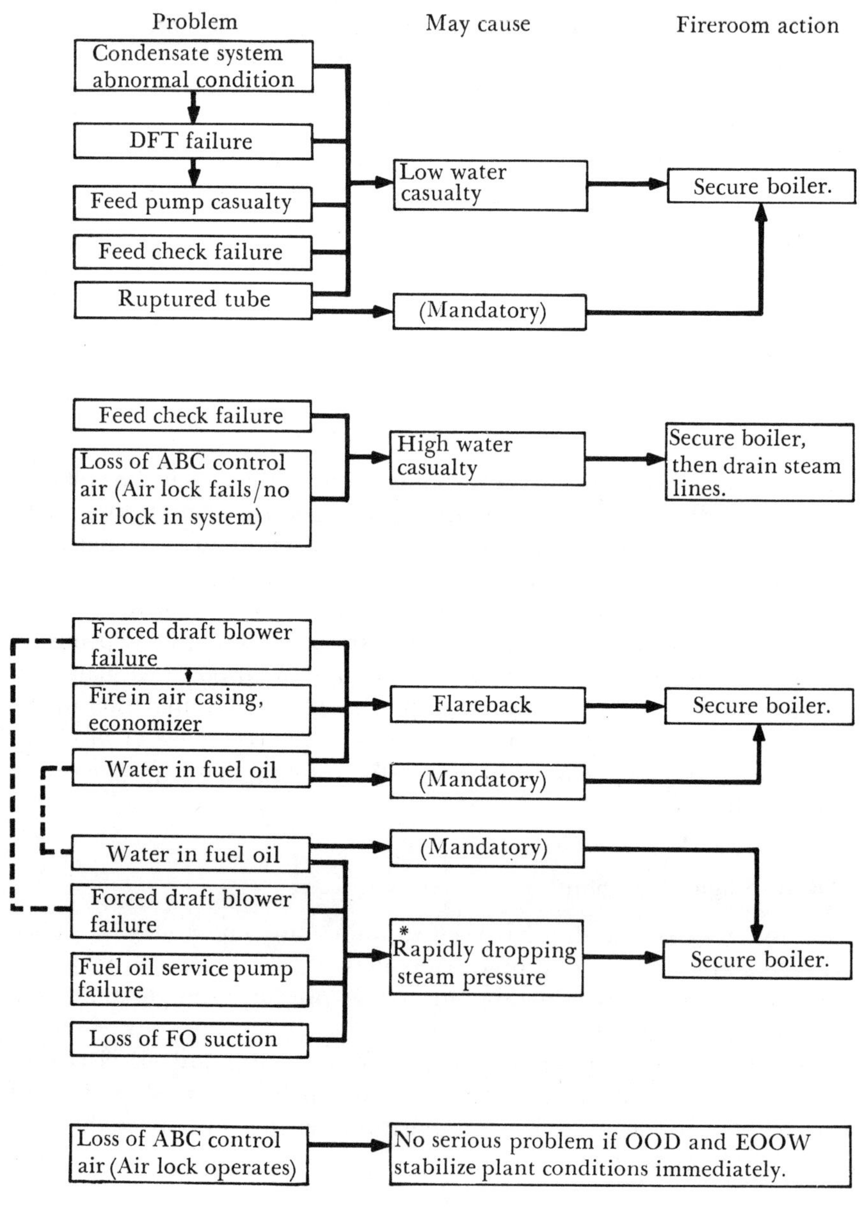

Table 3–1. Summary of fireroom casualties

* The term rapidly dropping steam pressure is arbitrarily defined, by the author, as an *abnormal* situation with potentially serious consequences and recognized as such by the EOOW or watch supervisor. As a further note, the OOD should follow the EOOW's recommendations concerning speed restrictions in *any* situation involving a dropping steam pressure, as long as such a recommendation would not otherwise hazard the safety of the ship.

(explained in Chapter 4) or a speed directed by the CO's policy. Inform the OTC of the casualty and maximum speed which can be maintained, as directed by the CO. Use the most secure method possible.

Cross-connection of main and auxiliary steam should be performed in a minimum amount of time if the casualty has been a low water, flareback, or rapidly dropping steam pressure.

If the casualty is the result of a ruptured tube or other pressure part, cross-connection may *not* be carried out until the EOOW has ensured that the boiler is completely secured or the pressure part isolated. Otherwise, the remaining boiler could be dragged off the line and men killed from escaping steam in the affected boiler/engineering space.

Cross-connection will be delayed somewhat after a high water casualty, since steam lines must first be drained to avoid damage to the main engine and turbogenerators.

CHAPTER REFERENCES

Boiler Technician First and Chief, NavTra 10536-E
NavShips *Technical Manual,* Chapters 221 and 079 (Vol. 3)
Principles of Naval Engineering, NavPers 10788-B

4

The Engineroom—Where It's Used

During the late 1960s a series of expensive and sometimes tragic propulsion plant accidents focused the attention of senior naval planners on the engineering reliability of surface warships of the U. S. Navy.

The majority of accidents occurred in the firerooms of destroyer and frigate type ships and in the steam generation areas of main machinery rooms in larger ships. Water quality and the proper operation of automatic boiler controls, two propulsion plant "insurance policies" examined in an earlier chapter, and other boiler-related problems came under close scrutiny. Money and manpower necessary for corrective action was allocated. Total propulsion plant reliability of the fleet began to improve.

About the same time, another development drew sharp comment from many senior engineers and dismayed numerous others. Significant numbers of single-plant, single-screw warships were either entering the fleet or under construction.

When the final 1200 PSI FF was commissioned, there were nearly 70 single-screw, escort type warships. (The *overall* total of single-screw warships includes high-value ships such as the helicopter carriers [LPH].) The number of single-plant, single-screw warships will be even greater in the 1980s with the addition of the gas-turbine-powered *Perry*-class guided-missile frigates.

The controversy concerning the single-screw warship centered around the fact that a single-plant ship is, inherently, not as reliable as a multi-plant ship. Early opponents of the single-screw warship recommended that towing at sea become a mandatory ship exercise. Proponents maintained that reliability would be achieved, and that the cheaper cost of a single-plant ship meant that more warships could be purchased to modernize a fleet that had suddenly grown quite old.

Both these developments in the late 1960s—the review of the fleet's propulsion reliability and the introduction into the fleet of numerous single-screw warships—placed engineroom operation into its proper perspective: Mobility is vital to a warship, and mobility is threatened by a malfunction of the main engine installation.

Indeed, the efficiency of a main engine, proper meshing of its turbines and reduction gearing, reliability of its lubricating system, and effectiveness of its condenser installation are of primary importance to any class warship. A general knowledge of engineroom operation, therefore, is important to the prospective OOD of any warship. An engineroom causalty on a single-screw warship may well leave that ship "dead in the water," at least for a while. In a twin-plant ship a main engine casualty could have a similar result if the "domino effect" were to occur.

In this chapter we will discuss in general terms a main engine installation, including the application of impulse turbines and reaction turbines to today's modern conventional propulsion plants. Additional topics include the following: the "astern" turbine; reduction gears, shafting, and propellers; the lubricating oil system; main condenser considerations and categories of engineroom casualties, including recommended actions by the OOD.

Before we begin, however, here is an update on the original controversy over the reliability of single-screw warships: The 1200 PSI FF/FFGs have proved to be highly reliable ships. As a class, their mobility achievements have largely silenced their original critics. Some of the doubts remain—but in a "hot" war any type of battle damage that would cause a frigate to lose propulsion would doubtless have some effect upon a multi-plant ship also. Meanwhile, the 1200 PSI FF/FFGs—with their two boilers *but* single main engine and single propeller—have been compiling an excellent mobility record.

Now it's time to consider propulsion steam turbines. As you recall from the economic analogy used in Chapter 1, turbines "buy" steam from the boiler and produce work.

PROPULSION STEAM TURBINES

General Turbine Information

The first steam turbine was built by Hero more than 2000 years ago. Men adapted turbines for shipboard propulsion about 50 years ago. Today steam turbines are used for propulsion in most surface ships and nuclear submarines. The following are some advantages of steam turbines over earlier methods of ship propulsion:

1. Steam turbines are capable of operating at high temperatures and pressures, and exhausting at very low exhaust pressures (high vacuum).

2. As a result of this temperature range, a turbine installation has a very high efficiency, requiring relatively little space and light foundations.

3. There are few vibrations because of few moving parts.

4. Both maintenance cost and consumption of lubricating oil are low.

Perhaps there once was a time when all that the Captain or engineer officer expected the OOD to know about turbines was a basic definition: a bladed wheel or *rotor*, contained within a *casing*, that turns when jets of steam move the *blading*. Now, however, the OOD or prospective OOD requires a great deal more information about turbines. The first item of information a prospective OOD should know about a turbine is its purpose: energy conversion.

Even on your first or second engineroom tour, you had an appreciation of the enormous amount of *thermal energy coverted to work* by the turbines when you noted the high temperature and pressure of steam admitted to the high pressure (HP) turbine compared to the relatively low temperature and pressure (high vacuum) existing in the main condenser. Turbines convert thermal energy into mechanical kinetic energy and then into work. *Velocity* exists wherever kinetic energy is present.

In turbines, as is true elsewhere throughout the steam cycle, temperature/pressure relationships are important. The *pressure drop* across the nozzle of an impulse turbine, or through a stage of fixed and moving blades in a reaction turbine, increases the velocity of steam passing that point. The steam is directed onto moving blades, and the higher the velocity of the steam, the faster the moving blades rotate.

Understanding the relationship of pressure drops to velocity, and velocity to the rotational speed of moving blading, is important to understanding how turbines operate. Subsequent subsections in this chapter will address the differences between impulse and reaction turbines and applications of each. But there is additional information about steam propulsion turbines in general that is more important to a prospective OOD than knowing how an impulse turbine differs from a reaction turbine. The following information is true for all types of propulsion turbines:

- Marine engineers strive for maximum efficiency in their design of axial-flow propulsion turbines.
- Numerous modern turbines are actually impulse-reaction turbines, with the first stage a velocity-compounded impulse stage (also called a Curtis stage), and succeeding blade rows using the reaction principle.
- The greater the pressure drop, the higher the velocity. Blade speed increases as velocity increases. To prevent damage from high centrifugal forces, steam must be expanded in stages.
- Pressure decreases in each stage because blades are longer and the casing diameter is correspondingly greater, allowing the steam to expand into a larger area than it occupied in the previous stage.
- Turbine tolerances are critical. Turbine *thrust bearings* absorb the

rotor's fore and aft axial movement. *Journal bearings* support the rotor and prevent radial misalignment.

• Leakage must be minimized. Each main engine installation has a *gland sealing steam* system which performs the following two functions: (1) prevents *air* from entering through the LP turbine glands and eventually destroying vacuum in the main condenser; and (2) prevents *steam* from leaking past the HP turbine glands and increasing ambient engineroom temperatures. *Labyrinth* and *carbon* packing have respective high pressure and low pressure applications in additional sealing of turbine glands at the ends of the turbine rotors.

• Turbines require *dry* and *clean* steam. A *steam strainer* is installed in the main steam line before steam can enter either the HP turbine (ship moving ahead) or the *astern* elements of the LP turbine (ship moving astern). The steam strainer is designed to prevent large slugs of water or foreign material from causing either blade erosion or damage.

• Turbines require uniform heating and cooling. A seriously *bowed rotor* will result whenever a hot turbine sits idle for any length of time. The LP turbine will be more seriously affected than the HP turbine, since the "cold" main condenser is directly beneath the LP turbine in most main engine installations.

Main Engine Warm Up and Cool Down/Main Engine Jacking Gear

Turbines require uniform heating and cooling. The *jacking gear,* also known as the *turning gear,* is a motor-driven gear mounted on the after end of the main reduction gear casing which meshes with the reduction gear when engaged to slowly turn the HP and LP turbine rotors during periods of warm up and cool down.

The jacking gear is also engaged to inspect the reduction gears and reduction pinions which compose the main reduction gear set, and to turn the main propulsion shaft the required number of turns during inport cold iron periods.

When the ship is making preparations to get under way, the machinist mates will bring the lubricating oil temperature to at least 90°F and request permission from the OOD (inport) or the command duty officer to engage jacking gear and begin to jack over main engine(s). (The engineers are also required to request permission form the OOD before engaging jacking gear to turn the propulsion shaft during inport periods or so the engineer officer can inspect the main reduction gear.) The OOD (inport) must *always* verify that the area around the propellers and shafts is clear before he grants permission for the engineers to jack over the main engine.

After sufficient vacuum has been established in the main condenser during underway preparations, the EOOW will request permission from the OOD to test main engines as a means of warming up the HP and LP turbines. When the bridge grants permission, the machinist mates will disengage the jacking gear and admit steam to the HP and LP turbines by cracking open first the ahead throttle and then the astern throttle. Opening the ahead throttle slightly warms up the HP turbine, while "astern steam" is more effective in warming up the LP turbine. The ahead and astern throttles are never opened simultaneously.

The jacking gear also has a locking feature which prevents movement of the propulsion shaft, reduction gearing, and HP and LP turbine rotors. Stopping a shaft for the purpose of engaging the jacking gear, or the jacking gear locking feature, is a precision operation *whenever* the shaft is turning. Placing an engine back in commission is also a precision operation. *Both* of these exacting evolutions are controlled by the EOOW. He must follow exacting procedures to prevent damage to vital propulsion machinery, and during these periods ship mobility is greatly restricted.

Even after an engine has been placed back in commission, speed of the shaft must be closely monitored to prevent damage to propulsion machinery from uneven heating. Improper warm up can cause excessive thermal stress and/or excessive vibrations from *bowed rotors.*

The jacking gear is also placed in operation after the ship moors or anchors in order to allow HP and LP turbines to cool down properly and to prevent condensation from forming inside the main reduction gear casing.

We paused briefly in our discussion of turbines in order to highlight the necessity of proper warm up and cool down of propulsion turbines. Proper warm up and cool down are equally important to reaction turbines and impulse turbines, which will be covered in the following subsections.

Impulse Turbines

In impulse turbines, the conversion of thermal to mechanical kinetic energy occurs in the *nozzles.* The shape of nozzles similar to those shown in Figure 4-1 causes an increase in the velocity of steam as it expands from a high pressure area to an area of lower pressure. This increase in velocity is accompanied by a corresponding increase in the kinetic energy of the steam. Nozzles also direct the steam flow against moving blades in an impulse turbine. (Stationary blades attached to the casing may also be used to direct steam flow in some impulse turbine designs.)

The velocity of the steam is reduced as the steam impinges upon the moving blades, and some of the kinetic energy is transferred to the moving blades. This results in a force that turns the turbine rotor. In an impulse turbine more than one row of moving blades may be associated with the nozzle set contained within the *nozzle diaphragm.* However, since the only

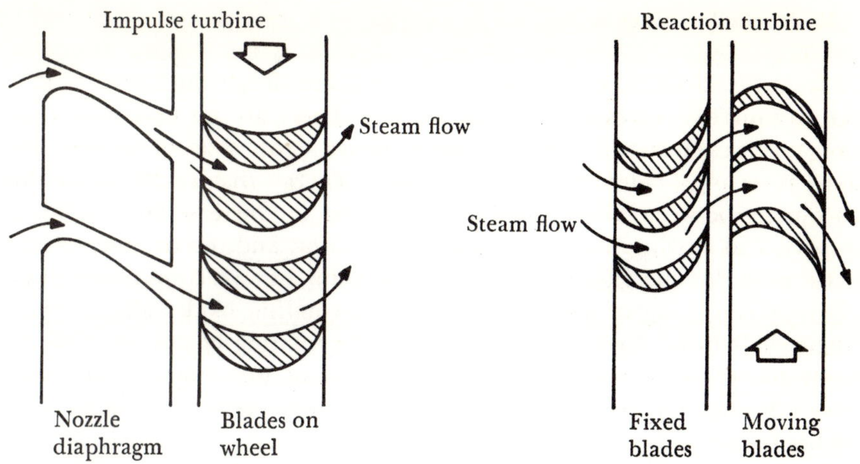

Figure 4–1. Impulse turbine nozzle and blading; reaction turbine blading

pressure decrease/velocity increase in an impulse turbine occurs in the nozzles, the number of stages of an impulse turbine is determined by the number of nozzle sets.

In theoretical terms, the rotational speed of impulse blading varies as one-half of steam velocity. What the turbine experts are saying in practical language is that impulse turbines can take the large pressure drops that result in high steam velocities and use these high velocities safely.

On board ship, this means that the HP turbine is an impulse turbine. Steam enters the HP turbine through a steam chest and flows along the axis of the turbine. The steam is expanded through successive stages, with each stage separated by a nozzle diaphragm. Impulse blading also permits partial admission of steam. In other words, the steam need not enter through all nozzles contained in the nozzle diaphragm.

The throttleman controls the amount of steam admitted to the HP turbine by manipulating the ahead throttle. Nozzle control valves lift in sequence according to cam action as the throttle is opened.

Periodically you may hear the engineer officer, main propulsion assistant (MPA), or senior machinist mate muttering exotic-sounding terms like "series flow" and "series-parallel flow." What they are talking about is the manner in which steam flows through their HP turbine to support various power demands. Steam flow through stages may be limited to only a certain number of nozzles in a nozzle diaphragm, and some stages may even be bypassed for varying speed demands. However, at full power steam flows through all the nozzles in all nozzle diaphragms to all stages within the HP turbine.

Reaction Turbines

In impulse turbines, the all-important pressure drop which results in a velocity increase takes place in the nozzles. There are no nozzles in reaction turbines. Instead, a pressure drop occurs across a set of moving and fixed reversing blades. Some of the kinetic energy that results from this encounter between a decreased pressure and an increase in velocity is imparted to moving blades attached to the rotor, and the turbine rotates.

Keeping this energy conversion in mind, think of a reaction turbine as a machine consisting of alternately fixed and rotating blades aligned on a single shaft in increasingly larger stages. The rotating blades are made to move the rotor by reaction to the expansion of inrushing steam which passes through one set of blades after another with resulting changes from higher to lower pressures.

In theoretical terms, the rotational speed of impulse blading varies as one-half of steam velocity. In general, blade speed of rotating blades of the reaction turbine varies directly with steam velocity. This allows LP turbines which operate on the reaction principle to extract energy remaining in steam, which exhausts into the LP turbine from the HP turbine via the *crossover line* with great efficiency.

Since the HP turbine of a warship is an impulse turbine, it would be convenient to say that the LP turbine is a reaction turbine in order to help differentiate between the two.

However, this is a "risky" generalization. Some manufacturers describe their LP turbines as impulse type turbines. Other manufacturers say that their LP turbines operate on the reaction principle. Certainly the blading differs, as is illustrated in Figure 4-1, and there are other important differences between impulse and reaction turbines. It may be more helpful for OODs and prospective OODs to think of the LP turbine in their ship as a turbine that combines the principles of both reaction and impulse turbines and exists in combination with the HP turbine to attain maximum propulsion plant efficiency.

Low pressure steam turbines used in the propulsion plants of surface warships are best described as *double-flow, impulse-reaction* turbines. (An illustration of this type of turbine is shown in Figure 4-2.) The LP turbine is physically larger than the HP turbine because larger blading and consequently larger stages are needed to convert energy remaining in the steam that has exhausted from the HP turbine. Steam from the HP turbine enters the LP turbine via the crossover line (previously mentioned) and expands through progressively larger stages toward either end of the LP turbine.

Prospective surface ship OODs should note that the main engine installation of some surface ships is composed not only of the HP and LP turbines, but may also include a *cruising* turbine. (The cruising turbine is a relatively small but efficient turbine used to make moderate speeds [ahead] in some classes of warships; cruising turbines are not used on board 1200 PSI warships.)

There are two additional highly important points to remember about LP turbines:

The first is that the main condenser is suspended from the bottom of the LP turbine in most propulsion systems, so condenser problems will almost invariably affect the LP turbine.

The second is that without the LP turbine in the propulsion plants of the warships of today, astern operation is impossible.

THE "ASTERN" TURBINE

The astern turbine, which permits the ship to move astern, is not really a separate turbine at all. Instead, the astern turbine consists of *astern elements* located at either end of the double-flow, impulse-reaction type turbine used as the LP turbine in the majority of conventionally powered surface warships. Figure 4-2 shows the upper half of this type of turbine with the astern elements in place.

The astern elements at either end of the LP turbine usually consist of two velocity-compounded impulse stages, also known as Curtis stages. *Baffles* protect both the astern elements and the final stages of the LP turbine ahead elements from steam exhausting directly upon either. Steam that remains after flowing through either the ahead elements or the astern elements, depending upon whether the respective bell is an ahead bell or a backing bell, exhausts directly into the main condenser.

When the bridge orders a backing bell, the throttleman closes his ahead throttle, opens his astern throttle, and steam is admitted to the astern elements at both ends of the LP turbine. On a back full bell or even a back ⅔ bell with only one boiler feeding the engine, an inexperienced throttleman who fails to watch his steam pressure gauge and opens his astern throttle too quickly suddenly finds the "restraining" hand of the engineroom top watch upon his shoulder. Figure 4-2 suggests why:

As the diagram shows, steam expands through only two stages when the ship is going astern. That's not very efficient compared to an ahead flank bell, when the steam is expanded through 16 or more stages in the HP-LP turbine combination of a typical 1200 PSI warship. The engineroom top watch knows—and quickly informs his throttleman in terse language!—that the massive steam demand required to supply the less efficient astern elements during a back full bell can quickly drag a single boiler off the line.

The following are some additional important facts about astern operation that the top watch wants his throttleman to know, just as the EOOW wants the OOD to realize:

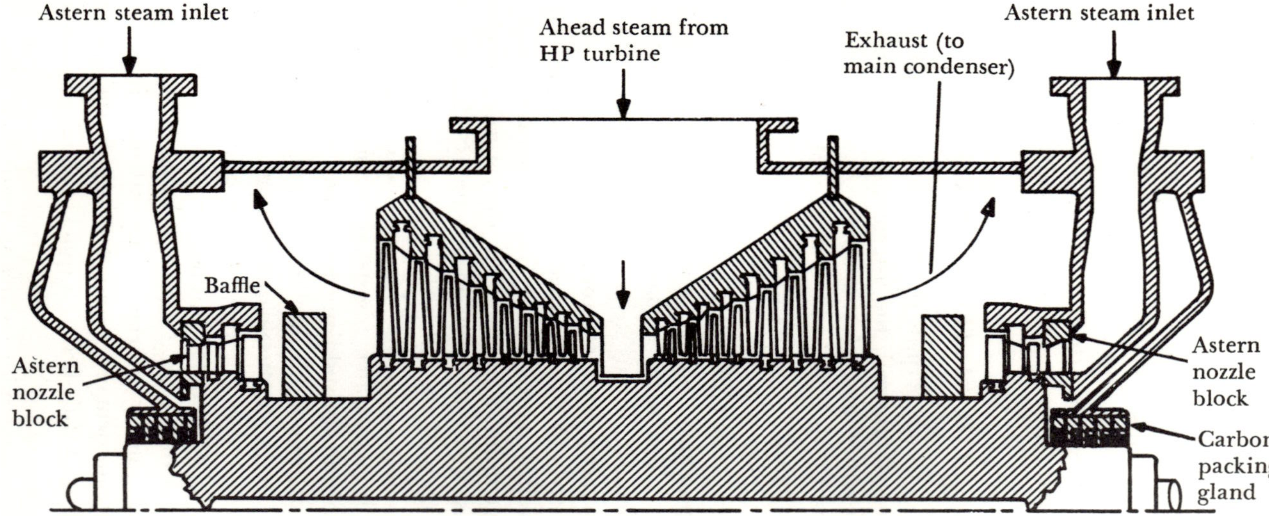

Figure 4–2. Impulse-reaction type turbine

• The maximum horsepower (measured by attaching a torsionmeter to the shaft during trials) that can be attained during full power astern is only about 15% of that which can be obtained during full power ahead in most warships.

• Backing down at high speeds for long periods is a critical operation that could cause serious damage to the main engine if key parameters are exceeded.

• The main circulating pump must be placed in operation whenever a backing bell is ordered, as well as during slow speeds ahead when *scoop injection* does not provide sufficient cooling seawater flow through the main condenser.

Concerning the maximum shaft horsepower that can be obtained when the ship is going astern, we have already seen that astern operation is much less efficient than ahead operation.

Consequently, the OOD, EOOW, and engineroom watchstanders all know that there are limiting RPMs for speeds astern as well as for speeds ahead. These limiting RPM vary from ship class to ship class.

In the case of prolonged backing at high speeds, the OOD, EOOW, and engineroom watch know that during astern operation the ahead elements of the LP turbine turn backwards. Heat from friction builds up in the ahead elements of the LP turbine until equilibrium is reached.

If the critical temperature is exceeded, the LP turbine casing will bend in an upward direction beyond acceptable limits. Serious damage to the LP turbine can result during prolonged backing at full power, especially if condenser vacuum decreases.

In addition to the LP turbine, the HP turbine may also be damaged during prolonged backing at full power. The crossover line between the turbines provides a path for heat that has built up in the LP turbine to enter the HP turbine. NavShips *Technical Manual* states that this process may be aggravated if the ahead throttle leaks through even slightly. (Unfortunately, this condition exists in many ships.)

During prolonged backing, the EOOW watches the main engine *exhaust trunk* temperature and other main engine operating parameters closely. The recommendations the OOD receives are based upon the EOOW's knowledge of these critical parameters and upon experience. A recommendation from the EOOW to reduce speed astern could come from the need to lower exhaust trunk temperatures, as well as to raise condenser vacuum.

As soon as the bridge orders a backing bell, the engineroom watch double-checks to ensure that the main circulating pump is pumping cooling water through the main condenser. If the "main circ pump" either malfunctions or is not put into operation, condenser vacuum will be destroyed and ship propulsion placed in jeopardy.

If the OOD indicates that the ship will be backing for a prolonged period, or moving alternately ahead and then astern for a lengthy period, the EOOW in a *controlled superheat plant* orders the fireroom to lower superheat temperatures.

In a 1200 PSI or other uncontrolled superheat plant, the EOOW requests the OOD to reduce speed astern as soon as critical temperatures are reached and/or a decrease in vacuum reaches a critical point.

SINGLE TURBINE OPERATION

The 1200 PSI FF/FFGs, which are single-screw warships, have the capability of operating with just one turbine if a casualty disables the other.

This casualty mode requires the use of special piping, blank flanges, and other special fittings. NavShips *Technical Manual* and the manufacturer's instructions specify the procedures to follow. Engineer officers of ships that have had to rig for casualty operation—and there have not been many—report that rigging takes from 24 to 36 hours.

A similar procedure had been developed for cargo ships produced during World War II.

If the HP turbine of a 1200 PSI FF is disabled, casualty operation using the LP turbine will enable the FF to proceed at a greatly reduced speed. Casualty operation using the HP turbine will enable the FF to go faster—but use of the HP turbine only, unlike use of the LP turbine in the casualty mode, does not allow the FF to back down.

REDUCTION GEARS, PROPELLERS, AND SHAFTING

Reduction Gears

Even at low power, propulsion turbines rotate relatively rapidly. In addition, the HP and LP turbines operate at different RPM (revolutions per minute). Propellers, however, must operate at much slower speeds for maximum efficiency.

The *main reduction gears* convert the rapid rotation of the HP turbine and the somewhat slower rotation of the LP turbine to speeds that allow the propeller to perform its work efficiently. Gear ratios of approximately 20:1 between maximum HP turbine RPM and maximum propeller RPM are common in many surface warships. To say it another way, the main reduction gears of a main engine installation allow both turbines and the propeller to operate within their most effective ranges of RPMs.

Reduction gears are used in diesel, gas turbine, and nuclear power propulsion plants to perform the same function that reduction gears do in the conventional steam plant—match the speed of the prime mover to the speed of the propeller. Reduction gears are also used in many different kinds of turbine-driven auxiliary machinery.

Mention main reduction gears to most engineering ratings and prospective OODs, and they visualize *locked-train, double-helical* (also called herringbone), *double-reduction* gears. One reason is that this type of reduction gear is part of each main engine installation in every CV, CG, steam-driven amphibious ship and auxiliary ship, DD, and FF type ship. Another reason is that almost every seagoing sailor or engineering classroom candidate has seen pictures of this type of reduction gear, or at least the outside of its casing.

The very phrase "locked train, double-helical, double-reduction" makes each installation sound intricate, efficient, and expensive. Each set of main reduction gears *is* intricate, efficient, and expensive. Moreover, there is but one set of reduction gears per shaft.

An explanation of the basic operation could be given here, but it's difficult to believe that an officer or senior petty officer of any rating doesn't understand the importance of the main reduction gears. Perhaps it's better to talk about the main reduction gears to these two men:

The engineer officer, who takes the *only* set of keys to the high-security padlocks from his safe on at least a quarterly basis, unlocks the inspection covers, and conducts the detailed inspection of the reduction gears required by the PMS.

The senior machinist mate chief petty officer, who scrutinizes and tugs at each high-security padlock and lead seal on each inspection and access cover before jacking over the main engine. A daily inspection of padlocks and lead seals on the reduction gears ensures that no malcontent has tried to damage the gears.

Propeller, Shaft, and Bearings

When you first began to study ship propulsion, you may have had difficulty understanding why ships move through the water. If Sir Isaac Newton had been nearby, he might have related the concept of action-reaction, as it applies to ship propulsion, somewhat like this:

The propeller is attached to the shaft, which is connected to reduction gears by the main shaft coupling. As the propeller rotates, it pushes against the water. But the water pushes back. The resultant force is transmitted along the shaft to the *main thrust bearing*.

The main thrust bearing is contained within the reduction gear casing on most ships. (The actual location of the thrust bearing within the casing depends upon manufacturer's design.) The main thrust bearing transmits the resultant force of the propeller vs. water encounter to the ship's hull.

Discounting such factors as propeller "slip," friction, and fouling of the underwater hull, the ship moves through the water because for every action there is an equal and opposite reaction.

That's the way Sir Isaac Newton *might* have explained ship propulsion, *if* he had been around at the time and *if* he had known that you or a shipmate were having problems understanding ship propulsion. Marine designers and the engineers who construct technical marvels like the main thrust bearing might wince at this simplification. Still, it may be reassuring to the beginner to be able to relate ship propulsion to Sir Isaac Newton and his Law of Action-Reaction.

Figure 4-3 is a simplified diagram showing functional relationships of ship propulsion (excludes generation, condensate, and feed phases):

Strut, stern tube, and line shaft bearings are shown in the figure. Although not as intricate as the main thrust bearing, these bearings also perform important functions by supporting the shaft and restricting radial play (movement).

The strut and stern tube bearings are lubricated by water. Line shaft bearings, also known as *spring bearings,* have their individual lubricating oil supplies.

Some spring bearings, like the stern tube bearings, are located in *shaft alleys* or normally unmanned spaces. On board many ships these unmanned spaces are locked when the ship is inport. The shaft alley is checked hourly under way and as often as required inport to ensure that excessive water is not leaking past that stern tube shaft seal and to verify that bearings are secure.

Damage to any bearing can quickly put a main engine out of commission and hazard the ship. Frequent inspections are required to ensure that bearing operation is normal and that no abnormal condition, such as a missing lubricating oil dipstick, threatens normal operation.

MAIN ENGINE LUBRICATION

All rotating machinery in the engineering department, including the emergency generators, requires lubrication. Individual *forced feed* lubricating systems protect high-speed, highly important equipments such as main engines, turbogenerators, emergency generator prime movers, main feed pumps, and, in most ships, the forced draft blowers.

With the exception of the main engine and reduction gear installation in FF-1040/FFG-1 class warships, the forced feed lube oil system of each main engine turbine and reduction gear installation includes *three* lube oil service pumps (LOSP). This highly reliable system includes the following pumps:

An *attached* pump, driven by the main shaft, provides sufficient oil pressure to the various turbine bearings and the reduction gears at higher speeds. The speed varies from ship to ship, but generally at about 12 knots the attached pump provides sufficient pressure to the entire system. (FF-1040/FFG-1 class warships do not have this attached pump, however.)

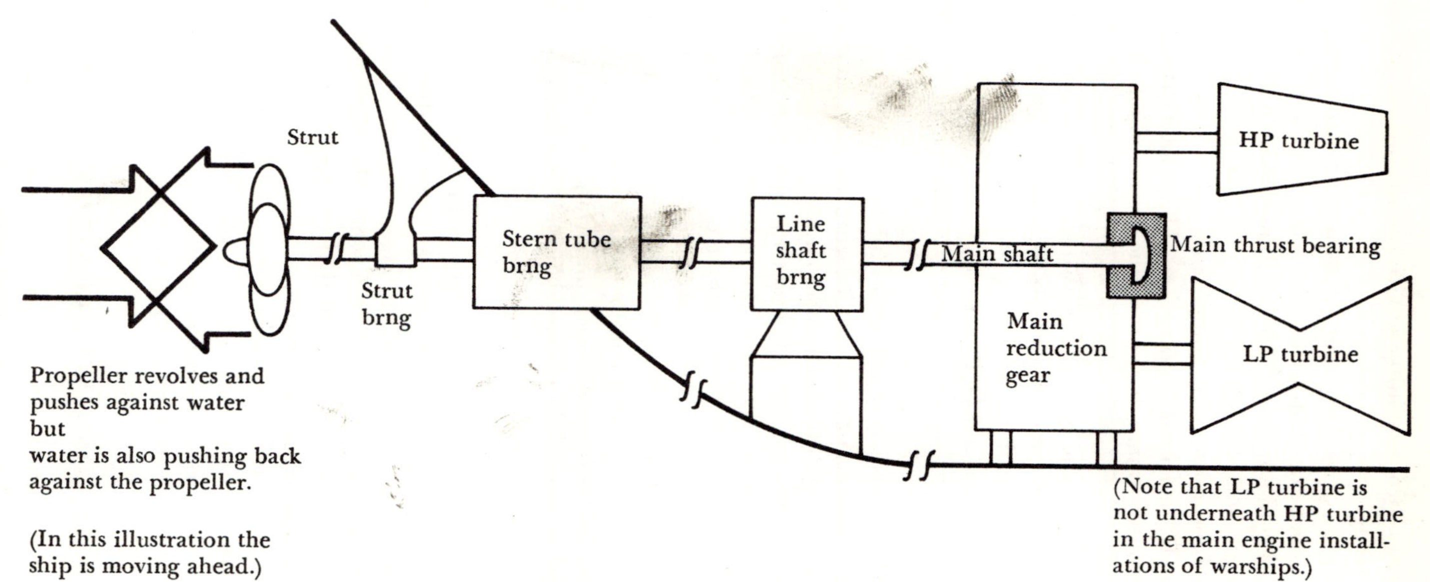

Figure 4–3. Simplified functional relationships of ship propulsion

If the ship slows, two remaining service pumps can act in succession to protect the main engine.

In aircraft carriers, cruisers, and destroyers, a *steam-driven lube oil service* pump assumes or shares the load when pressure supplied by the attached pump begins to drop. This turbine-driven pump uses 600 psi steam.

If the steam lube oil service pump fails, the *electric lube oil service* pump is set up for emergency operation. This motor-driven pump also serves as the standby pump if the turbine-driven pump is out of commission.

In the frigates and some other ships, two electric lube oil service pumps are used for lower speeds and to back up the attached pump. Each electric service pump has both high and low speeds. (FF-1040/FFG-1 class warships have two motor-driven LOSPs with these two speed settings.)

The pumps operate at various pressure settings specified by PMS. Individual Maintenance Requirement Cards (MRCs) also list the setting that activates the *lube oil alarm,* which senses lube oil pressure at the *farthest remote bearing.* The location of that bearing and the lube oil pressure required at that bearing differ from ship to ship.

After leaving the pump, the oil passes through the *strainer,* into the *lube oil cooler,* and finally arrives at the individual turbine bearings and reduction gear *spray nozzles.* Then the lube oil drains to the *sump,* where the pump(s) takes suction and the process continues.

An engineroom tour reveals that there actually is quite a bit more to the lube oil system. The system employs relief values, pressure-regulating valves, and lube oil heaters. Another important system component, the purifier, must be run *at least* 12 hours daily under way as specified by the Naval Sea Systems Command.

Basically, the lube oil service system is designed to deliver clean oil to the HP and LP turbine bearings and the reduction gears at 120–130°F. Reduction gear efficiency drops if the oil temperature drops below 120°F. Possible bearing damage may also occur.

Lubrication is so important that even when the engineroom crew engages the *jacking gear* to turn the shaft daily in port, oil must be supplied at a minimum temperature and pressure.

On board ship the *engineroom lower level watch* is an interesting person to talk to. He has a great deal of responsibility. He can describe a hot bearing, flooded condenser, and numerous other conditions that are rarely encountered but which can cause major casualties.

MAIN CONDENSER CONSIDERATIONS

Efficient turbine operation depends on steam with a high turbine inlet temperature and the low exhaust temperature associated with high vacuum in the condenser. When this designed temperature difference exists, the HP and LP turbine combination of a warship efficiently converts the thermal energy of steam into kinetic energy and then into work.

Turbine experts say that efficiency increases as much as 5% for each 100°F of superheat. Ships with 1200 PSI plants always have high superheat available, which is a definite advantage over other conventional steam plants.

Still, the propulsion plant of any conventionally powered surface warship requires a properly functioning main condenser to attain efficient operation. Nowhere in the propulsion plant is the critical temperature-pressure relationship, which is so highly stressed, more important than in the main condenser. The propulsion plant needs a low temperature area into which the turbines exhaust. The high vacuum in a condenser may be considered as a positive pressure so low that it approaches *zero psia* (pounds per square inch absolute); when the condenser develops this low positive pressure, the temperature in the condenser will also be low because of the temperature-pressure relationship. The low positive pressure the main condenser develops equates to high vacuum.

Designers strive for a condenser that can develop a vacuum in excess of 28 inches of mercury (28″ Hg). In some ships such as 1200 PSI CGs the vacuum may drop to about 25.5″ Hg under full power conditions, but, in general, main condensers of warships can achieve a vacuum in excess of 28″ Hg if there are no air leaks in the system and the temperature of the surrounding water is 70°F or less.

Air ejectors (there are two sets of main air ejectors for each main condenser *and* a separate steam reducing station to ensure that the air ejectors have sufficient steam to maintain their jet pump type evacuation of air from the condenser) create the initial vacuum in the main condenser, then help sustain this vacuum. During normal operation, however, the rapid condensation that occurs when steam exhausts from the LP turbine and comes in contact with cool seawater circulating through the seawater side of the condenser *tends to sustain* a greater share of the vacuum.

Throughout our school years we often heard the phrase: "Nature abhors a vacuum." Nature acts the same on board ship, and when vacuum begins to decrease, it can be difficult to regain. Worse yet, vacuum may even be lost completely. Here are some reasons why:

- *Insufficient cooling water flow:* The ship is going so slowly ahead or is backing down so that scoop injection does not operate, and the watch forgot to put the main circ pump into operation.
- *Decreased cooling water flow:* Perhaps the OOD decided to conn the ship through a big bed of kelp to look for sea life, or else mud is fouling the condenser seawater side as a result of maneuvering in muddy, shallow water. (Actually, cooling water flow through the condenser may be re-

duced in a number of other ways. The large number of plastic bags that a ship may encounter in the harbor of any major seaport throughout the world can clog condensers, as well as fire pump suction and the sea chests of other vital auxiliary machinery. OODs on board FF/FFGs backing down with Prairie Masker energized—more about Prairie Masker in a later chapter—also report that this class ship may suffer a drop in vacuum if the Masker air restricts cooling water flow.)

• *Air leaks* into the main condenser through LP turbine glands, or at some other location.

• *Loss of steam pressure* in the engineroom, which results in malfunction of the steam reducing station supplying the air ejector and the subsequent inability of the air ejector to help sustain condenser vacuum. Loss of gland sealing steam can also cause a loss of vacuum.

• *Insufficient main condensate pump discharge,* so that the air ejectors—which are cooled by the flow of condensate through them—overheat and thereby fail to help sustain condenser vacuum. (In addition, the fireroom watch will soon be complaining bitterly about the "lack of condensate" when the level of their DFT suddenly begins to drop!)

• In ships such as the 1200 PSI DD/DDG/CGs the *loss of makeup feed* (MUF) *suction*: In this situation the engineers have failed to shift makeup feed suction according to good engineering practice, and air rather than MUF is now being sucked into the main condenser. (This is not likely to occur in 1200 PSI FF/FFGs, since MUF enters the system via the fresh water drain collecting tank, where it is then pumped into the main condenser.)

As with other casualties, failure of the watch to recognize and react properly to abnormal conditions that will involve the main condenser can result in "dominoes" tumbling rapidly in the propulsion plant. For example, loss of vacuum can result in serious damage to both turbines and the condenser itself. This occurs when the increase of heat that accompanies loss of vacuum results in turbine misalignment. If heat continues to increase in the condenser steam side, rupture of the condenser expansion joint and/or tube sheet leaks may occur.

Although the requirements differ between ship classes, the rules concerning both *main circulating pump* and *condensate pump* operation are simple:

• The main circulating pump *must* be placed in operation whenever the ship is going astern and for speeds ahead that are insufficient to ensure adequate cooling flow through the main condenser. (This means that the engineroom watch in a 1200 PSI FF/FFG-1 will light off their main circ pump at 12 knots or less, while the engineroom watch on a 1200 PSI DD/DDG/CG will put their main circ pump in operation at 8 knots or less.)

• The second condensate pump *must* be placed in operation during

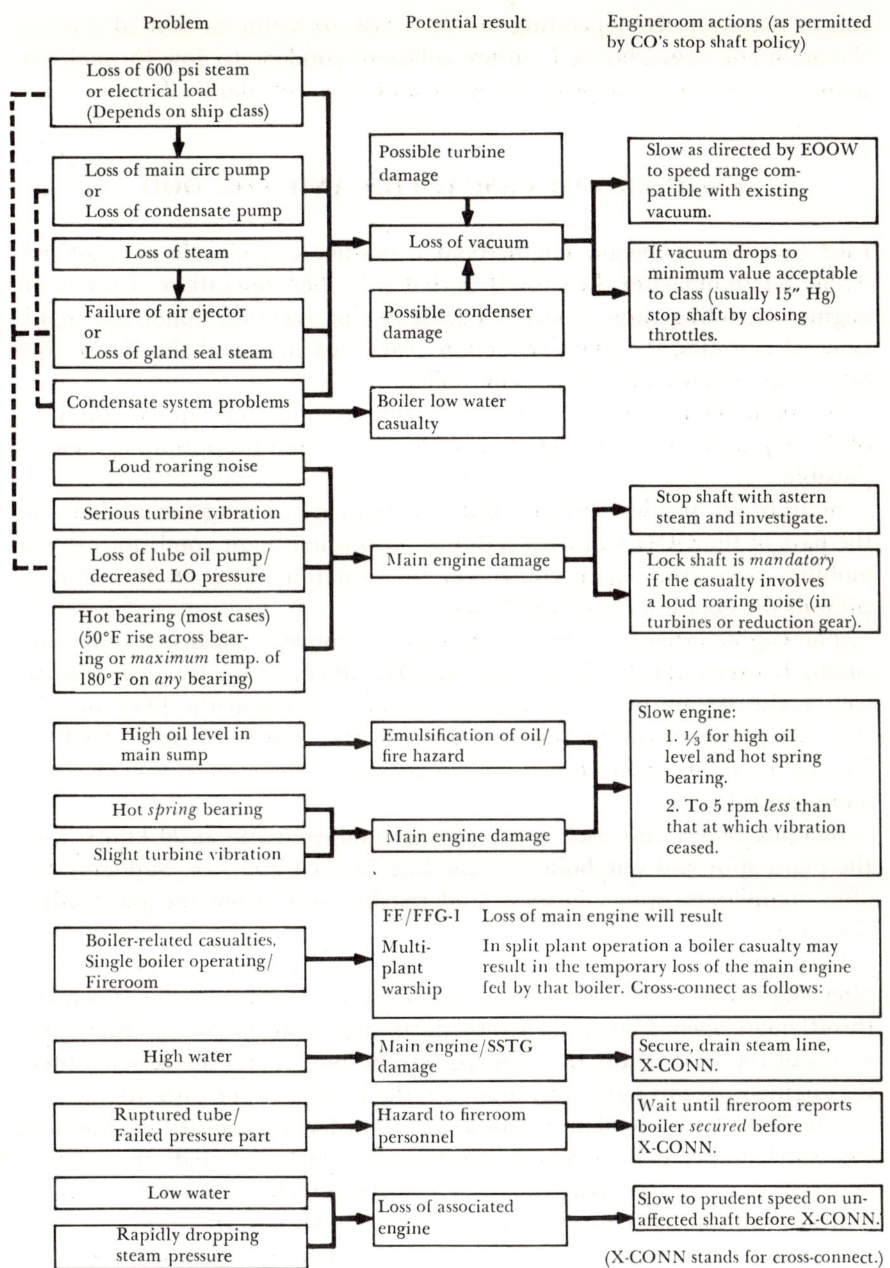

Table 4–1. Summary of typical engineroom/engineroom-related casualties

full or flank speeds, depending on ship class, to maintain desired level in the main condenser hotwell, ensure sufficient condensate flow through the main air ejectors for cooling purposes, and maintain the DFT water level.

ENGINEROOM CASUALTIES AND THE OOD

Like fireroom problems, engineroom casualties can also be categorized. Table 4-1 summarizes the casualties that will affect operation of the main engine, including those associated with loss of vacuum; abnormal conditions of turbines, the main reduction gears, bearings, and the lubricating oil system; and loss of a propulsion boiler.

In engineroom casualty control the same ground rules apply—the goal of the engineers is to act safely to restore services and propulsion as soon as possible.

In addition, prudent action—but not necessarily emergency action—on the part of the OOD will help stabilize propulsion plant conditions so the mobility team can regain control of the situation. The OOD *can* help prevent the dominoes from tumbling.

The engine order telegraph (EOT) is an important means of communication between the OOD and the EOOW during casualty situations as well as during routine steaming. In a casualty situation the EOT may be the *only* means of communication between the OOD and the EOOW, so let's see how this order-answer electrical repeater between the bridge and main control is used:

Imagine that a destroyer is proceeding independently at 20 knots, with the plant split and one boiler on the line in each fireroom. Suddenly the after engineroom watch hears a loud roaring noise from the port reduction gear.

Even before the engineroom top watch could inform the EOOW, the after engineroom throttleman began to close his ahead throttle. Then the throttleman opened his astern throttle to admit steam to the astern elements of the LP turbine in order to stop the port shaft. In the meantime, the engineroom top watch had informed the EOOW of the casualty.

The EOOW's immediate response was to order the forward engineroom top watch to come to standard speed on the starboard shaft in order to decrease drag on the port shaft. Then the EOOW moved his answer indicators on the EOT to *starboard—ahead standard; port—stop*.

The OOD acknowledged by having the lee helmsman position the EOT order indicators on *port—stop; starboard—ahead standard* momentarily, then having the lee helmsman position both the port and starboard EOT order indicators on *ahead standard*. By doing this, the OOD tells the EOOW that he wants to make standard speed on both shafts, even though this is not possible at present.

By this time the main control talker had reported the casualty to another sound-powered phone talker on the bridge (usually the lee helmsman via the 1JV or X1JV circuit), who relayed this message to the OOD: "Loud roaring noise in the port reduction gears. Proceeding to stop and lock the port shaft. Maximum speed available on starboard engine with the port shaft locked is (depends on ship class) knots."

As the OOD hurried to inform the Captain, his concern for the casualty to the port reduction gear was offset to a small degree by a feeling of relief that the starboard engine was still available for propulsion. Still, the OOD knew that he could not exceed the *maximum speed with shaft locked* for his class ship; during a wardroom seminar on propulsion a month earlier, the main propulsion assistant had made it quite clear that using the unaffected engine to exceed this speed restriction could overcome the locking feature of the main engine turning (jacking) gear presently locking the port shaft. If this happens, the MPA had explained, the reduction gear which had been locked (in this case the port MRG) would begin to turn again, enhancing the risk of damage to gears and pinions within.

(The OOD also reminded himself that he would have to follow the EOOW's direction when unlocking the shaft after such a casualty had been restored. Most EOOW's will request that the exact *RPM on the unaffected shaft* and *"astern" steam pressure required to stop the affected shaft* be duplicated when disengaging the jacking gear/jacking gear locking feature. In any event, they want to ensure that the torque on the locked shaft will not overcome the gear ratio between that shaft and its jacking gear. Should this occur, pieces of the jacking gear assembly may become bits of shrapnel, hurtling through the engineroom and threatening the safety of watchstanders.)

In the foregoing illustration, we assumed the destroyer was proceeding independently in deep, blue water during peacetime when the report came: "Loud roaring noise in the reduction gear!" Under such conditions, the policy of every commanding and engineer officer in the Navy is the same—"stop and lock," regardless of the type of warship. Most commanding officers agree that under these conditions the affected shaft(s) should be immediately stopped for various other engine casualties too, including *loss of lube oil pressure, serious turbine vibration,* or a *hot turbine bearing.*

But warships are warships, and even in peacetime each performs a number of operations routinely which could rapidly become perilous to that ship in case of a casualty to a main engine. Underway replenishment alongside another ship is one such operation. All ships must both leave and enter port. Towing is performed routinely as an exercise by numerous classes of ships, and the increasing use of helicopters in VERTREPs (vertical replenishment) has added to the number of coordinated operations between ships and aircraft.

As a result, the commanding officer of a warship of the U. S. Navy, together with his engineer officer, formulates a "Stop Shaft Policy," which is to be used on board *his* ship during the period of *his* command. The Stop Shaft Policy for a ship often varies from CO to CO. As an example of just one aspect of a Stop Shaft Policy, a former commanding officer of a cruiser, now a flag officer, required that his EOOWs be prepared to "Secure the Ahead Throttle to a main engine if a loud, roaring noise is heard in the reduction gear associated with that main engine when the ship is alongside another or involved in a similar situation involving restricted maneuvering." (The Captain knew this would give him those 8 to 10 precious seconds required to analyze a tight situation without committing him to a final course of action.) The final sentence of this aspect of Stop Shaft Policy read as follows: "Permission to open the Astern Throttle to stop the shaft *or* the Order to reopen the Ahead Throttle of this engine will be given only by the Bridge in such a situation."

In the immediately preceding example of a Stop Shaft Policy (for a loud roaring noise in the reduction gear), the commanding officer was inserting his own judgment as a factor in his policy. Such a policy risked damage to a main engine—but also minimized the possibility of even greater material damage to his ship *and* another ship, with a resulting reduction in total force readiness. However, another commanding officer might have preferred to keep mandatory stop and lock procedures in effect at all times for such a casualty.

Recommending a particular Stop Shaft Policy for any ship is beyond the scope of this book. Suffice it to say that every ship should have such a policy, and every OOD and EOOW must be thoroughly familiar with their Captain's Stop Shaft Policy for the period of his command.

Table 4–1 summarizes engineroom and engineroom-related casualties. The "Engineroom Actions" are recommendations only, and presume that the casualty described does not occur under conditions of highly restricted maneuvering.

The commanding officer of any naval ship requires that his line officers fully understand the Captain's desires concerning ship control safety before he qualifies any as OODs. This section, therefore, presents only recommended courses of action by the OOD during an engineroom casualty.

A similar section concerning fireroom casualties was presented in Chapter 3. It should be noted that boiler casualties almost always affect the engineroom, particularly when a ship is steaming with just one boiler per shaft or in a cross-connected status.

The remaining sections of this chapter assume that the ship described is under way with one boiler feeding its normally associated main engine. (This would be a split plant status for twin-plant warships.)

OOD—Single-Plant Ship

In ships of this type, which includes large numbers of FFs and the six FFG-1s, it's smart to realize that the EOOW doesn't have a periscope. You and he only have *one* plant, and a general knowledge of the tactical situation will help him perform as effectively as possible.

Exchange information with the EOOW as a matter of routine; then follow his recommendations during propulsion plant casualties. Unless you are in a situation that involves highly restricted maneuvering, his recommendations will generally parallel those listed as "Engineroom Actions" in Table 4–1.

If the ship is in formation and suffers a propulsion plant casualty, inform the OTC and ships nearby of your propulsion limitations by the most secure method possible. If complete loss of propulsion is imminent and your movements will not hazard your ship or others, position the ship the best possible way with regard to wind and sea.

Single-plant ships usually have at least one large diesel generator (SSDG) which is used as an emergency generator at sea. This diesel generator prevents the loss of vital services in case loss of propulsion is the result of a boiler casualty.

OOD—Twin-Plant Ship

Before beginning a discussion of actions that could be taken by the OOD of a twin-plant warship who is faced with loss of a main engine, it is beneficial to consider a propulsion plant procedure known as "Emergency Standard."

Emergency Standard is an engine order that allows one boiler to feed both shafts safely when cross-connecting after a casualty to the other boiler has occurred during split plant operation. The use of Emergency Standard is *not* an official policy of the Navy. The Fleet Training Group, Guantanamo Bay, which originated the term a number of years ago, recommends the use of Emergency Standard for the following reasons:

- If the OOD slows to emergency standard—which is 15 knots for DD/DDGs and CGs—after a boiler casualty that occurred during split plant operations at higher speeds, cross-connection can be made without the remaining boiler being dragged off the line because of the high steam demand that results from supplying both shafts from one boiler. When the plant is safely cross-connected, the OOD can ring up speed increases in small increments until the ship is making maximum speed at this reduced capability.
- If the OOD slows to emergency standard when a propulsion plant casualty occurs that requires stopping a shaft, he is assisting the EOOW.

Both men know that the astern elements of the LP turbine demand an enormous amount of steam to stop a shaft that has been making high speeds ahead. Slowing to emergency standard decreases drag on the affected shaft, which could prevent the engineroom watch from dragging their associated boiler off the line as their throttleman frantically opens his astern throttle to stop the shaft and prevent damage to the main engine.

The use of Emergency Standard is a recognized procedure in many propulsion plants, particularly in those of Atlantic Fleet DD/DDGs and CGs. A prospective OOD should find out if the use of Emergency Standard is a recognized propulsion plant procedure in his warship.

Even if Emergency Standard is not a propulsion plant procedure in a particular twin-screw warship, the engineer officer has specified the speed to which that ship should slow in case a casualty occurs that involves loss of a main engine or cross-connecting a boiler. The engineer officer may well specify a speed below 15 knots, particularly if the ship has not been at sea for some time and his new gang of engineers has had insufficient time to perfect the skills they require for casualty control. (If this is the case, the foresight of the engineer officer may well enable his engineers to gain control of a casualty quickly, rather than the casualty "controlling" them!)

Regardless of whether the speed is emergency standard—or 15, 12, or even 10 knots—the EOOW will ask for a speed that will enable him to cross-connect safely in case of a boiler casualty, or to minimize the possibility of damage in case of an engineroom casualty. Here again, if the OOD gives the EOOW a general idea of the tactical picture, both men can do their job efficiently.

Let's assume that a twin-plant warship suffers an engineroom casualty which involves stopping a shaft:

If the engineer officer and the EOOW decide no damage to the main engine will result from letting the shaft "windmill" through the water rather than engaging either the jacking gear or the jacking gear locking feature, the throttleman will close his astern throttle (as well as his ahead throttle). The report the OOD would receive in this case would be "shaft unlocked and dragging," followed by the maximum speed the ship could safely make on the unaffected engine.

The commanding officer will direct the OOD to inform the OTC of propulsion limitations by the most secure method possible.

Regardless of whether the shaft of the affected engine is left "unlocked and dragging," or locked, or the jacking gear is engaged, the twin-plant ship is now like a single-plant ship—only one main engine is available for propulsion. At this point the OOD of a twin-plant warship should begin to think like his counterpart on a single-screw ship: Maintaining propulsion with that single engine is vital, and should be foremost in his mind.

CHAPTER FOOTNOTE

Obviously, the engineroom contains much more equipment than that discussed in this chapter.

Many enginerooms contain both turbogenerators and distilling plants. Turbogenerator prime movers in more recently constructed ships, such as the 1200 PSI warships, have their own individual auxiliary plants, complete with condenser, circulating pump, condensate pump, and one set of air ejectors. (The turbogenerators of 1200 PSI 1052/78 class FFs are located in a main machinery space separate from the engineroom, however.)

The distilling units, called the "evaps" by many engineers, are vital to the entire ship, and these units are located in the enginerooms of many warships. So is an electric-driven fire pump and an HP air compressor, which is vital for systems such as the torpedo-tube charging system and the counter-recoil system of 5″/54 guns.

However, since this chapter attempts to deal only with propulsion topics, many subsystems have been omitted.

Also omitted is an explanation of how the EOT is used to indicate an *emergency bell.* This information is vital to a prospective OOD as well as to the OOD and JOOD; so it is recommended that each review the procedure for his *own* ship at frequent intervals.

One man referred to often in this chapter is the EOOW. In the following chapter we will examine his role in propulsion plant operations, as well as suggest ways in which the OOD and EOOW can complement each other as key members of the ship's mobility team.

CHAPTER REFERENCES

Encyclopaedia Britannica, "Turbines"

Modern Marine Engineers' Manual, Vol. I, Second Edition, Cornell Maritime Press

Principles of Naval Engineering, NavPers 10788-B

5

Propulsion Plant Partners

Thus far, this book has been primarily concerned with some of the "nuts and bolts" of conventional surface ship propulsion. In this chapter, however, we are more concerned with people than with principles of propulsion.

Certainly, the OOD should have some basic understanding of the natural laws that govern the operation of his propulsion plant, as well as a realization of why the plant responds to these principles. Men, however, *control* the propulsion plant. Chapters 3 and 4 described actions men take to control fireroom and engineroom casualties, respectively.

The engineer officer of the watch—the EOOW—is the man who exercises primary control of the propulsion plant on a watch-to-watch basis. While the EOOW "works" for the OOD, the EOOW also "works" for the engineer officer, who is naturally responsible to the commanding officer for the overall operation of the plant. Like the OOD, the EOOW must be certified in writing by the Captain before he is allowed to stand duty at his respective watchstation.

In practical terms, this results in the EOOW serving as both the primary advisor to the OOD concerning propulsion plant operation and as the man who is responsible for the operation of that plant on a watch-to-watch basis. The authority and responsibilities of the officer of the deck are specified in Navy Regulations. The EOOW is clearly subordinate to the OOD. What the Captain expects from both men is an effective partnership that results in the ship being able to perform her mobility missions safely and reliably. The OOD and the EOOW must speak a common language in their propulsion plant partnership.

The importance of an effective partnership between the OOD and the EOOW is the primary subject of this chapter.

The partnership and the relationship between the OOD and the EOOW consist of a number of concerns common to both men. This chapter will discuss these common concerns, which include the following: split plant operation, cross-connected operation, engineering conditions of readiness, critical periods, special evolutions, and economical operation.

Before beginning a discussion of these common concerns, however, it will be beneficial to review two major ingredients necessary to form an effective propulsion plant partnership: confidence and communication.

THE OOD AND EOOW—A MUTUAL CONFIDENCE

A major objective of this book is to give the OOD a review of basic propulsion plant operation. A secondary goal is to help acquaint a prospective OOD with what he needs to know about engineering to perform effectively as an OOD.

Accordingly, this book has stressed *basic* concepts, *basic* procedures, and *basic* equipment operation. To obtain his qualification, the EOOW has to know all this basic information, as well as a great deal more. Let's consider a main feed pump, for example:

An OOD who is comfortable with engineering knows the function of the main feed pump, the location of the pump, how the pump is powered, how many feed pumps are operationally available, the importance of the main feed booster pump, and what will happen if the main feed pump fails. The EOOW must know all of this information, too. In addition, he must have a thorough understanding of the following information concerning operation of the main feed pump: suction and discharge pressures, the protective recirculating feature, gland cooling system, lubrication system, safety features, valve lineup, and emergency operation.

The EOOW needs to know this additional information in order to perform his function of propulsion plant control. When one of the EOOW's primary assistants, the fireroom top watch, reports a problem associated with the main feed pump, the EOOW and the top watch (watch supervisor) must speak the same language. The EOOW also knows enough about the construction of the main feed pump to visualize the problem area.

Moreover, when the OOD calls main control and orders "Standby for 25 knots!", the EOOW will ensure that an additional main feed pump and the second condensate pump are placed on the line in the fireroom and engineroom (1200 PSI warships), respectively, in accordance with the engineer officer's standing orders.

The OOD can be confident that his EOOW will have a vast knowledge about main feed pumps, as well as other major equipments, systems, and propulsion plant procedures. The EOOW achieved his qualification through the Personnel Qualification Standard (PQS) for that watchstation, and all of the knowledge factors listed previously are required to obtain an EOOW qualification.

For the propulsion plant partnership to be truly effective, the EOOW similarly requires a certain degree of confidence in the OOD. An OOD who calls main control with the order to "Standby to make 25 knots," as an example, whenever time and the tactical situation permits, inspires confidence in his EOOW.

The reason for this is because a sudden and unannounced movement of the EOT from all ahead ⅔ to all ahead flank during a period of routine steaming naturally makes the EOOW wonder what will happen next. To make matters worse, if the EOOW calls the bridge to report that he will be unable to maintain such a speed because of a casualty to the second condensate pump and receives a response such as "So what does that mean?", both he *and* the OOD may be in for a very eventful watch together.

Confidence and communication are related. The following subsection deals with communications between the OOD and the EOOW. Before we discuss communication between these partners, however, consider once again the importance of confidence in this partnership. Confidence can enable a ship to maintain mobility in some situations when the ship might otherwise be forced to ring up "all stop." The following are two possible situations:

In the first, imagine that an FF-1052/78 class warship is encountering heavy seas while maintaining 10 knots. One lube oil pump is out of commission and has been CASREPT (reported as a casualty). Then the second pump suddenly fails.

The book solution calls for the shaft to be stopped immediately. But the EOOW knows the danger of the ship losing steerageway in heavy seas. So he informs the OOD but recommends also that the ship increase speed to 12 knots. He wants the attached pump to pick up the entire load so lube oil pressure can be maintained.

Now, at least, he and the OOD have borrowed time to plan their next step.

In the second example, the main circulating pump fails. But if the ship can increase speed so that scoop injection provides the required condenser cooling, the situation may still be saved.

The key point is that *men control the propulsion plant.* Situations arise that are not "covered by the book," but which can be resolved by men with proper expertise and mutual confidence exercising quick reaction.

THE OOD AND THE EOOW— THE NECESSITY OF COMMUNICATION

The importance of effective communications between these propulsion plant partners—and, indeed, all members of the mobility team—cannot be overemphasized. Perhaps the best way to stress the importance of effective communications, which in modern propulsion plant operations includes the use of the formal "repeat back" technique, is to recount an incident in which effective communications were completely lacking. This incident, also characterized by a lack of mutual confidence, really happened:

The aircraft carrier, a World War II veteran, was preparing to conduct routine flight operations in a non-combat environment. The Main Propulsion Assistant (MPA) was on watch in engineering central control as the EOOW. Aircraft had been "spotted" on the flight deck. Substitute the commanding officer of the CV for the OOD, and the "stage" is set. The rest of the story is related by the MPA:

"Down in Central Control we saw the rudder angle indicator swing right, and it looked as if the ship was ready for another afternoon of aircraft launches and recoveries. The next thing that happens is a ruptured tube in Number Four Boiler—and the next couple of hours I'll never forget!

"Fortunately, the watch evacuated that Machinery Room without any casualties. But they were having real problems securing the boiler with the remote-operated boiler stops and the remote fuel cutout valve. I wanted to cross-connect steam to that engine as soon as possible, but the report on getting the boiler secured seemed to take forever.

"The rupture was so massive that we found out later the furnace of the boiler was literally destroyed. Meanwhile pieces of economizer tubes and brickwork were being hurled up through the stack!

"Finally we get the boiler secured, and now I can start cross-connecting to the main engine in that space. Suddenly the Bridge wants to increase speed, and we almost lose another boiler by dragging it off the line while cross-connecting. . . .

"I told the Electrician of the Watch to strip non-vital circuits if necessary. We had already had to trip two turbogenerators off the line. Everything was happening so fast—but all these reports were going up to the Bridge. Matter of fact, one of the senior Chiefs on watch grabbed the phone from the regular talker and passed the word up himself so there'd be no mistaking how critical the situation was.

"Finally we get to the point where we're controlling the casualty, rather than casualties controlling us. That's when the Old Man's voice comes over the speaker—I'll never forget his words:

" 'I DON'T KNOW WHAT YOU PEOPLE ARE DOING DOWN THERE—BUT I WANT IT KNOCKED OFF RIGHT NOW!!

" 'YOU'VE GOT BRICKS AND CHUNKS OF METAL COMING UP OUT OF THE STACKS—AND IT'S GETTING ALL OVER *MY* AIRPLANES!!' "

Such an incident can have a humorous ending. (Notice, however, that we have made no mention of the expenditure of man-hours and money to repair damage to that boiler and, perhaps, the airplanes—and, such an expenditure is definitely not humorous.) But the odds are much higher that an incident such as this which involves a lack of communication will end in tragedy rather than humor. In any casualty, the OOD needs to be confident that his EOOW and the engineers are acting as quickly and competently as possible to restore services and to maintain mobility. The EOOW needs to have confidence that action by an OOD will not further endanger a ship struggling to control an engineering casualty.

Effective communications enhance mutual confidence. Naturally, the best way of controlling casualties is by preventing potential casualties and "tight situations" from occurring. But this is not always possible. Compounding the potential for real disaster is the fact that numerous engineering terms sound similar. Consider, for example, the excitement a bridge talker can create by mistakenly reporting a "(forced draft) blower casualty" as a "boiler casualty" at night as the ship plows through state four seas within 3000 yards of an aircraft carrier with the wind shrieking through the halyards.

Consequently, the OOD and the EOOW insist upon formal, declarative, and standardized communications approved by their respective type commander during propulsion plant operations. The term "MG," which we will see in Chapter 7 refers to a motor generator, can easily be confused with "TG," which we already know refers to a turbogenerator. Similarly, the terms "increase" and "decrease" sound much alike; so the EOOW would use the respective words "raise" and "lower" to report a change in condition. The formal "repeat back" technique is also a feature of Standard Propulsion Plant Communications Procedure, a standardized, written procedure for their ship that the OOD and EOOW must constantly review and use as partners.

Here are some other techniques successful OODs employ in establishing effective communications with their EOOWs:

• They get periodic updates on plant status from the EOOW when not much else going on is exciting except watching the big, blue ocean. That engineering status board on the bridge requires current information, in any event.

• They discuss special situations with the EOOW, such as the Captain's policy for a loss of lube oil pressure casualty alongside a replenishment ship. (The CO and engineer officer agree on the Stop Shaft Policy, and it may differ greatly from the standard stop shaft procedure.)

• They receive periodic reports on water consumption from the EOOW, especially during those periods when distilling capability is reduced.

• They keep the EOOW advised of the general tactical situation, especially those conditions that could require a greater degree of engineering responsiveness.

• They ensure that the JOOD and the boatswain's mate of the watch (BMOW) take active roles in bridge talker and propulsion plant talker training in Standard Propulsion Plant Communications Procedure, which is to be used for all steaming and casualty control conditions.

Establishing effective communications and building confidence between the OOD and his EOOW are, of course, common concerns to both. But now it's time to compare split plant with cross-connected operation, and to discuss other common concerns of the OOD and EOOW.

SPLIT PLANT OPERATION

Multi-plant, multi-screw, conventionally powered surface warships normally steam split plant. The easiest way to think of split plant operation is to consider a plant configuration in which one or more boilers supply steam to only one main engine, and usually one or two turbogenerators.

Split plant operation became standard practice during World War II for ships with more than one main engine for an obvious reason: Battle damage could knock out a key equipment in one plant, but the ship could continue to fight and move if only one other plant remained intact.

In fact, engineers of many ships are able to quarter their plants, so that loss of a main feed pump or even a fuel oil service pump results in the loss of only one boiler in a fireroom. "Quartering the plant" refers to plant lineup within the fireroom as a means of minimizing the possibility of loss of steam to a main engine when battle is imminent. In other cases quartering the plant may not be desirable. In any event, not all ships are capable of it.

System diagrams maintained at engineering control stations show the EOOW which valves he must keep closed to ensure that split plant status is maintained.

In the case of a destroyer or 1200 PSI cruiser there are *split plant valves* (SPVs) in both the forward engineroom and the after fireroom. Normal plant set-up calls for the forward boiler(s) to supply the starboard main engine and turbogenerator(s) in the forward engineroom, and the after boiler(s) to supply the port main engine and turbogenerator(s) in the after engineroom.

Figure 5-1, a partial diagram of the main steam system of a *Charles Adams*–class guided missile destroyer, shows the location of split plant valves in the main steam system of this typical 1200 PSI, twin-plant warship. Conventionally powered destroyers, guided missile destroyers of other classes, and the 1200 PSI CGs have a similar plant configuration. In Figure 5-1, valves MS 24 and MS 57 are the SPVs. The EOOW, who has spent many hours tracing the main steam system as well as other systems as part of his EOOW qualification, ensures that MS 24 and 57 are secured when the ship is operating split plant. He must also know the location of SPVs in the 600 and 150 PSI auxiliary steam systems, condensate system, auxiliary exhaust, and HP and LP drain systems.

(In multi-plant, 1200 PSI warships all systems with the exception of the feed system and 1200 PSI desuperheated steam to the forced draft blowers can be cross-connected by opening the respective split plant valves. The feed system cannot be cross-connected in 1200 PSI warships because of a design feature; if the EOOW must transfer feedwater from one plant to another during a casualty situation, he will direct that this be done with the use of the feed transfer pump.)

Many engineer officers of 1200 PSI DDs, DDGs, and CGs prefer that the main steam split plant valve in the after fireroom remain open. If a boiler casualty occurs, opening only the SPV in the forward engineroom—the location of main control in almost all steaming conditions—will quickly allow 1200 psi superheated steam to become available to all main feed pumps and turbogenerators, as well as both main engines, in the engineering plant.

While this practice is widely accepted, it is not always desirable. After prolonged use, valves may begin to *leak through,* and securing just one of the two SPVs may not, in fact, isolate a particular system. Major valves are often difficult and expensive to repair, and defective valves can be major headaches for the engineers.

The *condensate system* in particular can create sudden and dramatic problems in twin-plant, 1200 PSI warships if the system is not isolated by split plant valves in *both* the forward engineroom and the after fireroom. The problem stems from a situation in which the complete condensate system is located in the enginerooms of older destroyers and cruisers; in 1200 PSI DDs, DDGs, and CGs the boiler technicians and machinist mates share responsibility for the condensate system. Here's a brief description of the problem with the condensate system:

Condensate travels rapidly to an area of low pressure. Imagine that a DDG is operating split plant with the condensate systems in each plant isolated by just one SPV. If that valve leaks through even slightly, the DFT with the lower shell pressure will end up with an excess of condensate at the expense of the other DFT.

This situation can become even more critical during a casualty involving cross-connecting if the boiler technicians fail to completely isolate the DFT in the affected plant. Condensate will collect in the DFT in the fireroom in which the boiler casualty occurred. Meanwhile, the watch in the steaming fireroom will have difficulty maintaining water level in their DFT. They may have to resort to use of the feed transfer pump to maintain sufficient water in the DFT to feed the steaming boiler.

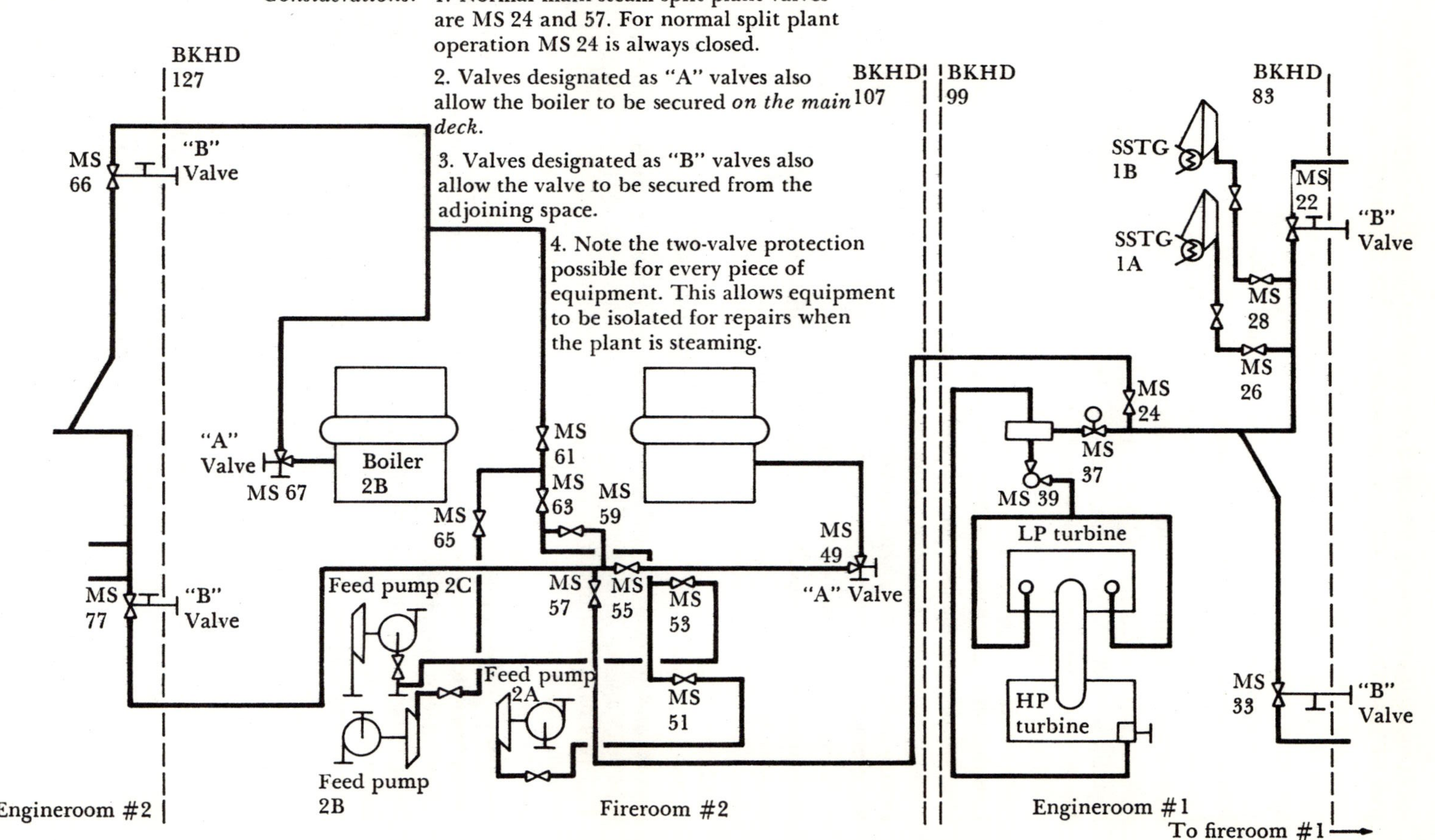

Figure 5–1. Partial diagram of main steam system—DDG-2 class

The condensate system can also cause major excitement after a boiler casualty has been restored, and the EOOW has ordered the plant split. Here again, improper valve lineup will result in lack of condensate in one plant and an excessive amount in the other. Communications can be interesting, with one fireroom accusing its associated engineroom of having put a second condensate pump on the line, while the other fireroom pleads frantically for more condensate.

Splitting the plant, as well as cross-connecting, requires that the EOOW *control* the situation. He must coordinate the actions of his watch supervisors—"top" watches—in their respective engineering spaces. Use of guidelines such as the Engineering Operational Sequencing System (EOSS) has standardized training for the engineers, but splitting the plant and cross-connecting are not "routine" evolutions. The EOOW, acting with the confidence of the OOD, is the man who controls the propulsion plant.

As a final note on split plant operation, a tour of the propulsion plant in many ships reveals valves candy-striped in a variety of colors. This does not mean the chief in charge of the space allows his men to express themselves artistically when painting valve handwheels. Instead, this provides the engineers with rapid identification of split plant valves for quick reaction during casualty control, especially in situations that require cross-connecting the plant.

CROSS-CONNECTED OPERATION

Officers of the deck in twin-screw warships normally think of cross-connected operation as an emergency situation resulting from a boiler casualty during split plant operation.

If a casualty required securing a boiler, the EOOW would order the plant to be cross-connected as soon as that could be done safely. He would cross-connect immediately in the case of a low water casualty, but cross-connecting after a high water casualty must be delayed until main steam lines to the main engines and turbogenerators have been drained to avoid thermal shock to turbines. As noted previously, there could be a considerable delay in cross-connecting if the boiler casualty involved a ruptured tube or other pressure part, since the EOOW must be positive that the boiler has been completely secured in such a casualty.

The following is an example of the report the OOD could expect in the case of a boiler casualty involving low water or a fire in the air casing, another type of boiler casualty:

"LOW WATER (FIRE IN THE AIR CASING), NUMBER 2A BOILER. SECURING NUMBER 2A BOILER. PROCEEDING TO CROSS-CONNECT THE PLANT. REQUEST YOU MAINTAIN

PRESENT SPEED IF SLOWER THAN 15 KNOTS (EMERGENCY STANDARD)."

This would be followed shortly by:

"THE PLANT IS CROSS-CONNECTED. NUMBER 1B BOILER AND NUMBERS 1 AND 2 MAIN ENGINES ON THE LINE. MAXIMUM SPEED AVAILABLE IS 21 KNOTS."

Of course, the OOD knows that he cannot immediately accelerate to 21 knots after having slowed to the speed which allowed the engineers to cross-connect his plant safely. (This was discussed in Chapter 4.) The EOOW, the OOD's partner, will inform the OOD of the estimated time to repair the casualty, the number of turbogenerators on the line, and other mutual concerns as soon as the EOOW knows that major systems including main steam, auxiliary steam, condensate, auxiliary exhaust, and HP drains are completely cross-connected.

Actually, cross-connected operation is not used solely as a means of maintaining propulsion after a boiler casualty. Many twin-screw surface warships periodically steam cross-connected as a matter of routine. The Captain may authorize cross-connected operation under the following conditions:

First, it may be used to conduct a bottom blowdown of a boiler as a means of maintaining water quality. Bottom blowdowns are required weekly by the Naval Sea Systems Command, and more often if total contamination rises above the limits specified in Chapter 2 (1300 and 700 micromhos/cm for 600 and 1200 PSI boilers, respectively). Normal procedure in many ships is to cross-connect the plant; secure the boiler requiring bottom blowdown; bottom blow that boiler after a proper cooldown period; relight fires under the boiler, bring it to line pressure, and open the boiler stop valves; and split the plant. A thorough bottom blowdown using this procedure takes about four hours.

In the second situation, the Captain may authorize cross-connected operation to allow the boiler technicians time to accomplish required fireroom maintenance. The BT rating is chronically undermanned, and the working hours of the BTs both at sea and inport exceed those of most other ratings. Cross-connected operation permits the BTs to catch up on maintenance during routine steaming periods.

The Captain has to balance his decision to permit cross-connected operation against operational requirements. The choice is not always an easy one.

For example, imagine that a DDG is en route to a missile firing exercise. The speed of advance (SOA) required to arrive on time is less than 16 knots, a speed which the DDG could make with one boiler supplying steam to both main engines in cross-connected operation. (At the time of this writing, fleet units are often restricted as a matter of policy to a maximum SOA of 16 knots while in transit as a means of conserving fuel.)

The Captain knows that his BTs could use this opportunity to accomplish fireroom maintenance during this transit. But the "weaponeers" work hard also, and loss of the electrical load as the result of a boiler casualty would negate the many hours they have spent "peaking and tweaking" their missile systems. In addition, the cost of replacing failed electronic parts that result from a sudden loss of the electrical generating capability of the ship can be tremendously expensive.

In this case, the Captain would probably conclude that the opportunity to fire one or more "birds" at a cost of $100,000 each cannot be jeopardized by steaming with a single boiler on the line in cross-connected operation. Weapons system readiness is as important to his ship as propulsion mobility; for a ship to be a fully effective combat system, *all* systems must be operational. In this case the Captain would probably insist upon steaming split plant.

If he chose to do so, however, the Captain of that DDG could authorize steaming one fireroom with a modified split plant configuration—and a maximum speed available of 27 knots—on the way to his missile exercise. This can be done in any 1200 PSI DD, DDG, or CG by steaming the after fireroom *only*. Figure 5-1 shows how the engineer officer and the EOOW would set up the plant for this propulsion configuration in their DDG.

The EOOW would isolate 2B boiler, 2B main feed pump, the port (number 2) main engine, and 2A and 2B turbogenerators by ordering main steam stops MS 63 and 77 secured. In other words, securing MS 63 and 77 would permit 2B boiler to supply steam only to 2B main feed pump and the after engineroom.

The EOOW permits 2A boiler to supply steam to 2A and 2C main feed pumps, the starboard engine, and both turbogenerators in the forward engineroom by opening MS 55, 57, and 24. (MS 24 and 57 are the 1200 PSI superheated steam SPVs in this class ship.) In this configuration the EOOW would also order MS 22 and 33 to be secured, which would isolate the forward fireroom for maintenance (or damage control) purposes. The EOOW would then have to ensure that specific valves in the 600 PSI, condensate, auxiliary exhaust, and HP and LP drain systems were positioned in such a manner as to "split" these systems also.

This is not the normal plant lineup. However, this particular lineup illustrates how the EOOW can control his plant with his extensive knowledge of the systems and plant diagrams. In addition, in some—but not all—warships the fuel oil service system within a fireroom can be split so that one pump takes suction from only one service tank and supplies fuel to only one boiler.

Still, the plant lineup just described is not a true split plant lineup. A casualty to the DFT in the after fireroom could affect both 2A and 2B boilers and, consequently, both main engines. Because of this we would have to conclude that the plant is still cross-connected.

If the DDG commanding officer authorized this particular modified steaming condition, the DDG would in one respect be operating with its propulsion plant similar to that of an FF or 1200 PSI FFG. There is but one DFT in the propulsion plant of these ship classes. Engineers in the 1200 PSI FF/FFGs report the only benefit that comes from operating with both boilers on the line is the reserve power that enables these ships to increase their maximum speed from about 21 knots with one boiler on the line to about 27 knots with both boilers on the line.

While the 1200 PSI DD/DDGs do not possess the emergency generator capacity afforded by the large and powerful 500 KW diesel generator (SSDG) in the 1200 PSI FF/FFGs, the propulsion reliability record compiled by the FF/FFGs would appear to make cross-connected steaming attractive for twin-plant warships under some conditions. If the tactical situation permits, savings in both fuel and man-hours can be realized by steaming cross-connected in a twin-plant warship at moderate speeds. Similarly, an FF/FFG-1 that can satisfy its speed and power requirements with just one boiler on the line is promoting fuel economy.

However, the tactical situation can change quickly. If a DD, DDG, or CG is operating cross-connected and a greater degree of engineering readiness is required, the OOD must give the EOOW as much advance notice as possible. Even in the case of an FF-1040 or FFG-1 class warship, which can have its second steam generator on the line in about 45 minutes in an emergency—or less than half the time it takes to put a conventional 1200 PSI boiler on the line—the EOOW needs as much advance notice as possible to be able to respond to an increased readiness condition.

The OOD and EOOW are propulsion plant partners who respond to the engineering condition of readiness prescribed by the commanding officer.

ENGINEERING CONDITIONS OF READINESS

The OOD and his EOOW must be speaking the same language when they talk about the condition of engineering readiness the Captain orders put into effect.

Many younger officers think the term "condition of readiness" applies only to progressive stages of sensor and weapon readiness. So when the senior watch officer fixes one with a piercing look during a wardroom training session and asks, "What's our Readiness Condition Two?", the answer can be interesting.

Sometimes the young officer gulps, gropes, and then goes into total recall: "All directors manned, CIC in port and starboard, 5″/54 carrier rooms manned," or whatever Readiness Condition Two might be for a particular ship.

But recall is not always total. Sometimes the officer forgets to mention engineering response.

This is not too surprising, however, because the status of the engineering plant required in Condition Two and Three is not defined for many ships. As a result the Captain, engineer officer, OOD, and EOOW speak of engineering readiness in conventional surface ships in terms of speed capability, generator capacity, and propulsion plant survivability.

Others may define engineering readiness more simply as "boiler power available."

From the previous discussions of split plant and cross-connected operation as each related to propulsion of a twin-screw warship, we know that split plant operation gives the plant increased reliability and survivability. Condition One is well-defined for all ships: *all* systems either on the line or ready for instantaneous use. Condition One for the engineering plant (in other than drill situations) requires that full boiler power be available. The plant is split in such a manner that maximum survivability for that class ship is achieved. A twin-screw warship might steam cross-connected as a matter of routine in Condition Four (depending on the Captain's desires), but all higher conditions of readiness require split plant operation.

Regardless of whether any particular condition of readiness addresses both sensor/weapons system readiness *and* engineering response, a higher readiness condition means that more men must man watchstations. The ship is, in reality, a combat system. The engineering plant and the engineers are vital elements of this combat system. Men who man the stations supporting a combat system lose effectiveness under prolonged conditions of maximum readiness.

The manpower requirements placed on the engineering department may differ from those of other departments during varying conditions of readiness. Consider this example:

A carrier is racing into position to launch strikes against designated targets. Opposition is expected, which may come from anywhere: missile-firing surface ships, submarines, or aircraft.

As the carrier nears launch points compatible with the combat radii of its aircraft, the task force commander orders increased weapons system readiness. But to sustain the speed necessary to arrive at those launching points, the engineers on board the carrier, her escorting surface missile ships, destroyers, and frigates all may already have been at "General Quarters" for much longer than 24 hours.

Any full-boiler-power, high-speed operation for prolonged periods involves trade-offs in fuel, manpower, and equipment. Fuel is limited. Men tire. Equipment may fail.

In order to conserve his engineering resources and to enhance his plant's responsiveness—and to strike a balance between the two—the Captain may order an engineering readiness condition put into effect that differs from the readiness posture of the rest of the ship. Readiness posture changes according to the tactical situation.

The OOD and the EOOW must know which engineering readiness condition is in effect, and the capabilities and limitations of their propulsion plant for each condition.

Engineering readiness, a common concern of the OOD and the EOOW, is increased to support many special evolutions, such as underway replenishments or entering and leaving port. Many special evolutions, which are common concerns of the OOD and the EOOW, too, are also critical periods of propulsion plant operation. The OOD and the EOOW must know what evolutions and periods are critical in propulsion plant operation. Critical periods, therefore, concern both the OOD and the EOOW.

CRITICAL PERIODS

Experienced officers and petty officers say that there is no such thing as a noncritical period in propulsion plant operation. They are correct.

Still the following are only a few of the peacetime situations that represent *highly* critical periods in propulsion plant operation: leaving and entering port, underway replenishments, heavy weather steaming, operations with aircraft, formation maneuvering, towing, and anchoring.

Destroyer and frigate sailors would add antisubmarine warfare exercises and "plane guard" assignments with an aircraft carrier to this list. Sailors on board amphibious ships would include operations involving the landing of the Landing Force.

The officer of the deck and his EOOW on board *any* ship engaging in *any* of these operations recognize that the particular evolution is a critical period requiring propulsion plant reliability. The EOOW and his top watches are particularly vigilant during these periods to ensure that mobility and services can be maintained in all situations. These men, along with the OOD, review casualty control procedures to ensure that they can respond to an engineering casualty.

Two periods, however, are particularly dangerous because engineering problems are more apt to occur then than at other times. The engineer officer of your ship can tell you which periods if you haven't already guessed: those first hours under way and those last hours before the ship returns.

Even though the engineering department reported "Ready for getting under way" to the executive officer along with the other departments, senior engineers know their plant is now in a particularly vulnerable period. The propulsion plant will be even more vulnerable if the ship has been inport for a long time, although the engineers lighted off their plant and steamed "auxiliary" for a few days before getting under way.

The engineer officer knows that some watchstanders may be new at their stations, and even an excellent PQS program cannot completely compensate for lack of practical experience. Engineering supervisors know which equipment and systems have had maintenance either performed or deferred. Numerous items associated with the engineering plant must be checked in the first hours under way.

So the engineer officer and his principal assistants are—or should be—on the deckplates during this period. Their flashlight beams probe into corners. They listen for changes in the whine of the auxiliary turbines and vibrations of the main engines. They recheck oil level in the spring bearings. They inspect logs to ensure that entries are being made correctly and to find out how much makeup feedwater the plant is using. They monitor temperature rises throughout the plant as steam demand increases.

On the bridge, the OOD finds time to review casualty control procedures, even though his watch is indeed busy. The OOD, his partner the EOOW, and the entire mobility team succeed because of their increased watchfulness.

But during the second highly critical period—those last hours at sea before the ship arrives inport—even the OOD and the EOOW may fall victim to "end-of-cruise complacency".

Perhaps the engineroom lower level watch notices the bearings of the auxiliary condensate pump feel hot—why bother to light off the other turbogenerator and its auxiliary plant right now? Or perhaps the low pressure air compressor supplying control air needs overhaul, the control air system is contaminated, and the ABCs have become sluggish—won't the ABCs still answer bells well enough to get the ship safely alongside the pier?

If the OOD and EOOW must fight complacency, it's a safe bet that other members of their watch team must also.

This kind of complacency is a shortcoming that has claimed many victims. And it's the reason why experienced engineers say there is no such thing as a noncritical period in propulsion plant operation.

SPECIAL EVOLUTIONS

The OOD and EOOW need to coordinate propulsion plant participation in special evolutions, another common concern of these propulsion plant partners.

In many cases these special evolutions involve maneuvering considerations. For example, leaving and entering port, underway replenishments (UNREP), and heavy weather steaming, as well as the other evolutions listed in the previous section as critical periods, all involve maneuvering considerations.

Evolutions such as these require that the propulsion plant be as reliable and responsive as possible. In the case of a twin-screw warship, the OOD must ensure that his EOOW is advised of the event in sufficient time so that the plant can be split for maximum reliability. This advance notice also enables the EOOW to have steering and other vital systems checked if this is included as one of his responsibilities in preparing for an evolution such as an UNREP. In the case of a single-plant warship such as an FF, advance notice will enable the EOOW to have the second boiler on the line in time for the UNREP or for departure or entry into port.

During any special evolution involving maneuvering, the EOOW *must* control all aspects of propulsion plant operation. As strange as it sounds, accidents and near-accidents have occurred because auxiliary machinery was shifted during a special evolution *without* the EOOW's permission. (A normal procedure in some ships consists of shifting from one propulsion auxiliary equipment to a standby unit at a specified time each day, such as the midwatch. You can imagine how your ship could be jeopardized by such a practice if an equipment malfunction occurred while shifting condensate pumps, as an example, while your ship was alongside another in an UNREP during the midwatch!)

The EOOW also must control all aspects of propulsion plant operation during Quiet Ship Operation, a program that is receiving increased emphasis in the U. S. Navy:

(If you ask the EOOW what the Quiet Ship Bill means to him, and he replies in a hushed voice, "That's sonarman stuff," or "Is that when everyone walks around without shoes and keeps from dropping wrenches on the deckplates?", the Quiet Ship Bill needs more emphasis in your ship!)

Unclassified instructions issued by type commanders emphasize that ships operating below cavitation speeds using a specific equipment configuration, with that machinery in proper repair, reduce their radiated noise significantly.

Reducing radiated noise allows ASW ships to improve their detection capability in both active and passive sonar search modes. It also means that non-ASW ships may run quietly enough to hamper an enemy submarine in its efforts to classify them—this, of course, gives friendly ASW forces a greater opportunity to detect the enemy submarine.

The overall noise reduction program incorporated into the construction of the 1200 PSI FF/FFGs—and improved to an even greater degree in construction of the gas turbine–powered warships—has made the Navy's newer ASW ships extremely quiet ships capable of exploiting their high-powered sonars. The 1200 PSI FF/FFGs, gas turbine–powered *Spruance*-

class DDs, and gas turbine-powered *Perry*-class FFGs also have Prairie Masker, which further reduces the radiated noise signatures of these ship classes.

Prairie (propeller) Masker (machinery) functions by emitting large quantities of air through tiny holes in the propeller blades and other surfaces near the propeller as well as through Masker belts located along the underwater hull. Prairie air prevents cavitation from occurring at low and moderate propeller speeds. Machinery and hull noise is insulated by air emitted through the Masker belts. Prairie Masker air in 1200 PSI FF/FFGs is supplied by large air compressors powered by 1200-psi desuperheated steam in FF-1052/78 class warships and 600-psi auxiliary steam in FF-1040/FFG-1 class warships. Prairie Masker air is "bled" from the propulsion gas turbines in the *Spruance*- and *Perry*-class warships.

With regard to the fleet-wide Quiet Ship program, conventionally powered surface ships radiate the least noise when they operate with the minimum equipment necessary for reliable mobility at the slowest speed that can be sustained with scoop injection. Of course, First Division cannot be chipping the deck, the engineers cannot be dewatering their spaces with eductors, and everyone may have to forgo a daily change of clothes since the laundry may not be operating.

Executing the Quiet Ship Bill requires coordination, supervision, and practice. Each ship has a Quiet Ship Bill which prescribes various conditions of quietness. The OOD and EOOW play major roles in implementing the Quiet Ship Bill, as well as many other special evolutions.

ECONOMICAL OPERATION

The final concern of the OOD and EOOW to be discussed in this chapter is economical operation. Both men first want to ensure that their plant is in the appropriate readiness condition required by the tactical situation. After the readiness requirements are satisfied, both the OOD and the EOOW want their plant to be operated as economically as possible.

Since the EOOW controls the propulsion plant, he is also the man who has primary responsibility for operating the plant economically.

If the OOD is truly interested in ways in which the EOOW strives for propulsion plant economy, the two men will find themselves discussing such topics as "condensate depression" and "motor-driven auxiliaries vs. turbine-driven auxiliaries."

Condensate Depression

Condensate depression, the OOD will discover, doesn't mean that the engineer officer or the EOOW is unhappy with the condensate system.

Instead, condensate depression refers to the difference between the temperature of the condensate discharge and the condensing temperature, which corresponds to the vacuum maintained at the exhaust steam inlet to the condenser (that familiar and important temperature-pressure relationship, once again!). An efficient condenser is one in which condensate depression is maintained at a reasonably low value under normal operating conditions.

If excessive condensate depression exists, plant efficiency suffers because the subcooled condensate must have additional heat added when it becomes feedwater/boilerwater. This involves an increase in fuel usage. In addition, excessive condensate depression is accompanied by increased absorption of air by the condensate. This air must be removed to prevent oxygen corrosion in the boiler watersides.

The EOOW attempts to keep condensate depression within allowable limits during steady state steaming by ordering the engineroom watch to throttle the overboard discharge of circulating seawater, which serves as the cooling medium in the main and auxiliary condensers. Condensate depression can normally be maintained within allowable limits if the temperature of the condenser overboard (seawater) discharge is kept about 10°F higher than the seawater injection temperature.

Turbine-Driven vs. Motor-Driven Auxiliary Equipment

When the two men discuss turbine-driven vs. motor-driven propulsion support equipment, the OOD will probably discover that the EOOW prefers to use motor-driven pumps to enhance economy if a choice exists. Here are some of the reasons the EOOW would give:

1. The use of electric pumps conserves steam and feedwater.
2. The plant uses less fuel because of the first benefit listed above.
3. The prime mover is simpler to place into operation.
4. Reduced maintenance is associated with the use of electric-driven pumps.
5. Adequate back-up power for motor-driven pumps is available in modern warships if the primary source is lost.

The EOOW would also add that turbogenerators (SSTGs) should be loaded to at least 70% of capacity, according to the Naval Sea Systems Command—so why not use motor-driven pumps rather than turbine-driven pumps if a choice exists?

If the OOD has exchanged information with his EOOW, however, the EOOW—the man who controls the propulsion plant—may decide the tactical situation requires the use of turbine-driven pumps to support the propulsion plant. The reason for this is that turbines do not depend on electricity at all. Even though major improvements have been made in circuit breakers, high temperature insulation, voltage protection, and other reliability factors for electric motors, turbine-driven pumps are still more dependable than motor-driven pumps.

Still, design improvements in motors, controllers, and distribution systems of the more modern warships have resulted in greater use of motor-driven auxiliary equipment on board these ships. Many warships, in fact, now have only motor-driven pumps to support the steam cycle. But where a choice exists, many EOOWs prefer to steam with turbine-driven pumps on the line if the tactical situation requires a greater degree of readiness.

THE EOOW–THE MAN WHO CONTROLS THE PLANT

This chapter has been devoted to exploring common concerns of the OOD and his EOOW, as well as emphasizing the mutual confidence and communication that must exist between these propulsion plant partners. However, little has been said about the EOOW himself.

The EOOW on board a destroyer or frigate will often be a chief petty officer, usually a boiler technician or machinist mate in the case of a conventionally powered warship, and an engineman/gas turbine specialist (GS) in the gas turbine–powered warships. In larger ships the EOOW will often be a commissioned officer. Regardless of the EOOW's rank or rate, he is a man who understands and values teamwork. Consider these remarks of a master chief machinist mate, a man who qualified as EOOW on board various classes of amphibious ships, destroyers, and cruisers:

"I can't think of a tougher job aboard ship than that of the OOD. He's up there day and night, in fair and foul weather.

"I'll never understand how he tells what all those lights mean at night, and which way he may have to maneuver the ship to avoid a torpedo attack. Even if the watch is not very busy, just let him miss a radio message at the same time that the Old Man walks on the bridge, and things can get pretty exciting pretty fast!

"The OOD has tremendous responsibility. If he doesn't do his job right, especially when our ship is nearby another one or in heavy seas, all of us could be in trouble. He's got a tough job—I want to help him do it right."

In order to qualify for his watchstation, the EOOW must be certified in writing by the Captain after completing the PQS for EOOW. A true "professional," the EOOW understands that the prospective officer of the deck or other young officer may have difficulty at first comprehending even such basic concepts as the response time of the main engines. He can explain the relationship of the Standard Acceleration and Deceleration Tables to fuel economy, as well as discuss engineering subjects with a complexity beyond the scope of this book.

Like many of us, the engineer officers of the watch in your ship may have been a bit bewildered on first descending into an engineering space in a ship. But now your EOOW is a highly knowledgeable man willing to share his knowledge with you—and this information could help you become more comfortable with engineering.

The engineer officer of the watch is a man with the extremely important job of controlling the propulsion plant. The watch he stands is of such importance that present Bureau of Personnel policy concerning the EOOW watchstation is as follows:

In order to become qualified as a surface warfare officer, an officer must qualify as a junior engineer officer of the watch (JEOOW).

In order to become qualified for assignment as executive officer or commanding officer of a surface warship, a surface warfare officer must qualify as engineer officer of the watch in that class warship.

CHAPTER REFERENCES

Principles of Naval Engineering, NavPers 10788-B

6

The OOD Inport—In Hot Water During Cold Iron?

Despite the old saying that "Sailors belong on ships, and ships belong at sea," a warship actually spends a considerable amount of time inport. The first four chapters of this book were concerned with building a basis for understanding how the steam propulsion plant operates, and Chapter 5 dealt with the partnership necessary to control this propulsion plant. In this chapter we'll stay "inport" for a while, and examine some of the considerations of the OOD about his engineering plant when his ship is not at sea.

A ship that is not under way will be in either one of two basic engineering conditions. In the first, the ship will be operating various equipments in order to provide its own electricity, steam, and firemain pressure. The second condition is commonly known as "cold iron," a term used to describe the engineering plant of a ship which is receiving some or all of its vital services from an outside source.

Warships that are at anchor or moored to a buoy generally provide their own electricity, steam, and firemain pressure. The OOD of a ship moored to a buoy or anchored is often posted on the bridge, unless his ship is in an extended upkeep/liberty status or in a liberty port. His propulsion plant partner, the EOOW, is on watch in main control. The OOD should know how much time is required for the EOOW to have the propulsion plant ready to get under way and to "answer all bells." The OOD needs to keep the EOOW advised of any changes in underway time and any other situations, such as major weapons systems light off for PMS, that might require the EOOW to put additional equipment on the line.

Ships that are moored alongside a pier or nested with other ships may also provide their own services, just as if each were anchored or moored to a buoy. In most cases, however, a ship moored alongside a pier or nested with other ships will be in a cold iron status. The OOD of a ship in cold iron will be stationed on the quarterdeck, but the status of his engineering plant is still a matter of major importance to him.

(The OOD inport, who may be a lieutenant in a ship such as a CV, or an ensign or senior petty officer from any of the shipboard ratings in a smaller ship, is referred to as the "quarterdeck watch officer" by some type commanders. In this book the title of OOD inport refers to the man who has the responsibilities and title of "quarterdeck watch officer.")

Generally, the OOD inport does not possess the basic engineering expertise required of his counterpart who stands watches on the bridge. This chapter, then, is devoted primarily to the OOD inport, the man who stands his watches on the quarterdeck. Regardless of whether the ship is in a cold iron status or providing its own services, the OOD inport needs a basic understanding of the following engineering topics:

Cold iron status
Auxiliary steaming/preparing to get under way
The firemain system
The ship service electrical plant inport
Security of engineering spaces
Fuel handling
Response time

Before we discuss cold iron, however, it should be noted that the OOD inport—as well as the OOD under way—has assistants to keep him advised on engineering status, even during cold iron periods.

All ships have a sounding and security watch to monitor conditions in engineering spaces and other areas required by a particular command. In smaller ships the sounding and security watch, who must be qualified in writing, makes his report to the OOD on the quarterdeck on a regular basis.

In many ships, including DDGs, *Spruance*-class destroyers, CGs, and all other larger ships, the damage control central watch serves as the focal point for all engineering watches. The DC central watch can rapidly provide the OOD on the quarterdeck with the following information:

• The status of the firemain and the number of pumps required to maintain pressure.

• The status of the ship service electrical distribution system, and whether the diesel generator is set up for *automatic start* in the case of the 1200 PSI FF/FFGs or in *lock out,* which protects the prime movers of the emergency generators in older ships from possible damage if shore power is lost momentarily during cold iron. (This will be explained in greater detail in a later subsection of this chapter and in Chapter 7.)

• Which spaces are under surveillance from roving or on-station watches, and whether spaces such as shaft alleys and steering gear rooms are locked.

• The location of major repair work in *any* part of the ship, particularly on jobs requiring fire watches.

All ships also have a duty engineering officer, who can provide information on plant status or other engineering topics that the OOD inport may desire. Similarly, each ship also has a Ship's Information Book (SIB). The SIB should be consulted to provide additional information for a particular ship on topics that will be discussed in this chapter.

COLD IRON STATUS

Extended periods of "cold iron" are more common for U. S. Navy warships now than they were 20 or even 10 years ago. Rising fuel costs have reduced the annual average number of days under way for many ships. Additional savings in fuel can be achieved by providing electricity and steam from the pier for ships moored alongside, rather than requiring each ship to provide its own electricity and steam.

Although maintaining ships in a cold iron status reduces the Navy's total fuel bill significantly, many senior engineers in conventionally powered surface ships prefer to light off and "steam auxiliary" on a regular basis. They point out that lighting off to provide steam to the turbogenerators to furnish ship's (electrical) power benefits the engineers' training and ensures system operability, especially during extended periods of cold iron. The engineering supervisors add that the routine light offs make the transition from furnishing steam for ship's power to furnishing steam to the main engine(s) a systematic evolution. (Introduction of the Engineering Operational Sequencing System—EOSS—to the fleet began in 1973. EOSS standardizes engineering training and provides step-by-step, systematic procedures for going from cold iron to "ready to answer all bells" and back to cold iron.) In situations where (electrical) shore power is limited, the weaponeers appreciate the engineers' going on ship's power to allow their shipmates to perform required PMS on today's complex combat systems.

Still, men on board ship appreciate cold iron, that period when their ship receives some or all of her vital services from the pier or another ship. In addition to saving fuel, periods of cold iron are valuable to the following men for these reasons:

• The engineers, who get the opportunity to perform required maintenance on their propulsion equipment, a vital requirement for any ship but especially for a single-screw warship. (Watchstanding requirements for the engineers are also reduced during cold iron.)

• The radiomen, who get the opportunity to go aloft to clean antennas and to replace insulators near stacks, so communications will remain reliable during the next underway period.

• The boatswain's mates, who get the "green light" to perform required topside maintenance, without the engineers heralding the appearance of fresh gray paint topside with numerous specks of fresh black carbon.

Many OODs inport also enjoy periods of engineering cold iron. Those molecules of H_2O are no longer racing through the boiler, main steam lines, and the turbine of a turbogenerator (SSTG) before exhausting into a condenser. And while electrons still speed through the ship service electrical distribution system, nevertheless it seems somewhat more peaceful because the electrical generating source is either ashore or on board another ship.

"It's quiet down in the plant," the OOD says to himself. "We're on cold iron."

But at this point, it's a good time for the OOD on the quarterdeck to consider what else cold iron means, as follows:

• Far fewer men are on watch in the engineering spaces.

• The ship still needs firemain pressure for a variety of reasons, including the cooling of equipment such as low pressure air compressors for ship service air, and for flushing water throughout the ship (except in *Spruance*- and *Perry*-class warships).

• Unless firemain pressure is being provided from the pier or another ship—and most engineer officers greatly prefer to isolate their firemain to their own ship—at least one electric fire pump in either an engineroom or a pumproom must be on the line to provide adequate firemain pressure.

• Various parts of the ship service electrical distribution system are energized to provide power to that fire pump and other important users of electricity throughout the ship.

• Steam from a source outside the ship passes through various small lines at reduced pressures for use in the galley and laundry and to heat water in hot water heaters throughout the ship.

• Water continues to collect in the bilges of *any* ship, unless the ship is in drydock.

All of the preceding situations can create problems for the OOD inport. He may suddenly find himself in the proverbial hot water, even though his ship is in cold iron.

Of course, the engineering plant of a ship steaming alongside the pier or in a protected anchorage may give the OOD inport a great deal of anxiety, too. Events—some with unfortunate consequences—can take place rapidly under almost any condition. But more about auxiliary steaming later. A casualty occurring during cold iron can provide all the excitement anyone needs. Here's just one example:

The cruiser was in a cold iron status, moored alongside a destroyer tender for the final two weeks before beginning an extended deployment.

Approximately 10 days remained before the engineers would light off and "steam auxiliary" to ensure that the ship would be in a fully combat-ready status when the deployment began.

Relatively few major equipments were required to support the cold iron status: two electric fire and flushing pumps (hereafter referred to simply as "fire pumps") and a low pressure air compressor. The LP air compressor —one was located in each fireroom in that class ship—was needed to supply a constant purge for the automatic boiler controls and to provide sufficient air, after further purification, for the waveguides of the ship's expensive and high-powered air search and missile fire control radars.

Normally, cooling water for that LP air compressor and other equipment located in either fireroom of that class ship is provided by auxiliary salt water cooling pumps located in either fireroom. The auxiliary salt water cooling pumps in this case were out of commission, and both had been CASREPT (reported as being out of commission in the Casualty Reporting System).

Still, the engineer officer considered that to be no major problem, since cooling water *could* be supplied from the firemain at a reduced pressure by means of a reducing station located in either engineroom. So cooling water was in fact being supplied from an engineroom firemain/cooling main reducing station, a common practice in many ships and the normal lineup for cooling water in 600 PSI destroyers.

So far, no problem. The night might have remained calm and peaceful alongside the destroyer tender too, except for a number of circumstances. These included—among other factors—rising bilge levels in the main engineering spaces; the requirement to comply with environmental regulations concerning discharge of bilge water inport; a lamentable scarcity of floating oily water/bilge water containers (called "doughnuts," because of their shape) for the ships moored in that port; and a realistic desire of the engineers to protect their equipment, particularly motor-driven auxiliaries, from the rising bilge levels.

Two other ingredients necessary for the casualty to occur were also present. The first was an equipment malfunction. The second was personnel error. Here's how each contributed to the casualty:

The engineroom cold iron watch (one man was posted as a cold iron watch for both enginerooms, and a second man was posted as the cold iron watch for both firerooms) had been using an *eductor* (a water jet pump on the *main drainage system*) to lower the level of the engineroom bilges as necessary to prevent damage to equipment. Equipment on the lower level of the engineroom, which was susceptible to flooding, included a motor-driven main condensate pump, motor-driven auxiliary circulating pumps and condensate pumps for the turbogenerators, sections of the static exciters for the turbogenerators, various motor-driven pumps associated with the distilling unit in that space, and the motor of the fire pump in that engineroom. The engineer officer had authorized use of the eductor to dewater main engineering spaces when bilge levels became dangerously high, a measure that would preserve engineering readiness without circumventing environmental regulations.

In this case, it didn't preserve engineering readiness.

After lowering the bilge level to a safe point and securing the *eductor actuating valve* to prevent backflooding into the engineroom through the eductor, the engineroom cold iron watch continued on his rounds. However, fluctuations in the firemain pressure, which occur when an eductor is put into operation, triggered a malfunction of the firemain/cooling main reducing station relief valve.

(The function of a relief valve is to protect a system or equipment from damage caused by excessive pressures. The firemain/cooling main reducing station relief valve protects the cooling main from excessive pressures. In the 16 and 26 class cruisers, these particular relief valves divert water to the engineroom bilges rather than over the side of the ship to protect the cooling main.)

When the relief valve malfunctioned, the valve failed in the wide open position. The engineroom began to flood quickly as seawater from the firemain was diverted into the bilge. The engineroom cold iron watch returned a short time later. But he gave only a cursory glance at the engineroom lower level from the top of the ladder before departing to wake his watch relief. Meanwhile, the relief valve—which had had all required maintenance performed on it since a similar malfunction occurred 18 months previously—remained in the wide open position. Water continued to pour into the forward engineroom.

Personnel error now became a major factor when the oncoming and offgoing watches did not conduct a joint inspection of the enginerooms before the watch was relieved. Then, when he noticed the water was now above the lower level deckplates in the forward engineroom, the man who had just taken the watch ran back to ask his shipmate for an explanation and to wake the duty machinist mate rather than taking immediate action to stop the flooding himself.

By the time the duty MM arrived, the water was approximately 12 inches above the engineroom lower level deckplates. All motors below that level were now immersed in saltwater. Flooding ceased when the duty MM quickly secured electrical power to the lower level, tripped the reducing station relief valve, and put the engineroom eductor into operation.

Decisive action by that well-trained petty officer prevented further damage. But he and other members of that work center, as well as the electrician's mates, were required to work many extra hours to repair equipment so the CG could begin her deployment on schedule. Two of the men working the long hours included the offgoing and oncoming watchstanders whose negligence contributed to the flooding casualty.

However, the negligence of those two men was not the full extent of the personnel error that contributed to the casualty. (And I ought to know: I was engineer officer of that cruiser.)

An investigation established that the major cause of the flooding was the malfunction of the firemain/cooling main reducing station relief valve. The investigation also established that neither man had completed his PQS for cold iron watch, and interim PQS qualifications had not been issued. Both men believed their training to be deficient.

In addition to being remiss about the qualification of cold iron watch-standers in my department, I failed to act on my belief that a watch-stander should have been posted in each main engineering space with the ship that near to its upcoming deployment. The number of men on watch when the casualty occurred in fact exceeded the type commander's requirement at that time. But the overall result of having two men on watch rather than four ultimately proved detrimental to morale, in this case, when many men had to work extra hours to compensate for an organizational weakness. The ship sailed as scheduled, but the man-hour expenditure required to meet the sailing date was high.

Some contemporaries have similar tales of grief. The engineer officer of a similar class cruiser experienced two floodings during his tour. Both floodings were the result of a similar malfunction of the firemain/cooling reducing station relief valve. Each time the engineroom was flooded to the upper level deckplates.

The preceding example illustrates how flooding can occur in a particular class ship. At the time of this writing (1978) the potential for a similar flooding because of a reducing station relief valve malfunction exists in a number of class ships. These include, but are not limited to, almost all of the 1200 PSI CGs and DDG/DDs. Flooding in these ships and many other classes of warship can also occur if the *eductor actuating valve* is not secured properly after bilge eductors have been used to dewater a space.

During one two-year period, 1972–74, the Navy averaged at least one serious flooding *every* month, according to unclassified messages from the various type commanders. Many of these floodings occurred during periods of cold iron. In some cases ships missed commitments, or were late for commitments because of a flooding or fire that occurred during cold iron.

Flooding alarms have been installed in the main engineering spaces of many ships in recent years. Increased emphasis has been placed upon PQS, including All Hands Damage Control PQS (which each man must complete within six months of reporting to the ship to which he is presently assigned). Additional "doughnuts" have been procured, although the number often seems inadequate for the ships moored inport. These improvements have helped reduce the number of floodings, but a serious flooding during cold iron is always a distinct possibility.

If a flooding (or fire) occurs, the OOD inport will not be the man who manipulates a valve or lights off a piece of damage control equipment to attack the casualty. But the OOD should remember that his ship is particularly vulnerable during cold iron.

The knowledge the OOD has of systems such as main and secondary drainage systems and portable and installed firefighting equipment—which he can obtain from his All Hands DC PQS—may enable him to help direct efforts to control a casualty during cold iron.

The OOD inport should know what constitutes cold iron on board his ship, as well as the watch organization that is monitoring the condition of cold iron in the engineering plant. If a casualty occurs, the OOD inport may be required to decide quickly that the ship should be called to General Quarters if sufficient manpower is unavailable to deal with a casualty that occurs during cold iron.

AUXILIARY STEAMING/PREPARING TO GET UNDER WAY

Even though many surface warships spend considerable periods on cold iron, sooner or later the time comes when the engineers light off their plant. Perhaps the ship will sail within a few hours. Perhaps the Captain and the engineer officer may only want to "steam auxiliary," as mentioned previously, to conduct engineering training, and to give the weaponeers the opportunity to "peak and tweak" the combat systems using ship's electrical power.

A young officer, chief personnelman, sonar technician first class, or any other OOD inport may find himself with the quarterdeck watch during engineering plant light off or auxiliary steaming. Let's assume that he knows the possible perils of cold iron.

He may find that lighting off or auxiliary steaming is even more perilous. Here's what happened on board a DDG of the U. S. Navy moored in Toulon, France, during the late 1960s:

The time was shortly before 0700, a "ho-hum" time of the morning. The engineers were conducting "routine" light off in the forward plant. Suddenly an explosion shook the ship.

According to reports, the force of the explosion was so great that unsecured deckplates (more about these in Chapter 9) were hurled up and out of the fireroom. Intense heat blistered paint topside in the vicinity of the forward fireroom and the forward stack.

Whether General Quarters was ever sounded is unknown. In any event, damage control central was not manned. The ship's inport emergency repair party arrived on the scene expeditiously. So did the Captain and the executive officer and the engineer officer—and many men in the repair party said later that at some time or other each was taking orders from either the CO, the exec, or the chief engineer as well as the repair party leader.

Fuel could not be secured topside because the master quick closing valve was inoperative. (Figure 3–2 shows the location of the master fuel oil QCV in a simplified fuel oil service system.) Investigating officers later concluded that the fuel oil service pump could have been secured by someone reentering the fireroom lower level through the escape trunk. But when the heat is so intense and reports are so unclear, you don't always think about those things.

French firemen succeeded in putting out the fire nearly three hours later. The cause of the explosion was undetermined. Fuel that could not be prevented from entering the boiler firebox accounted for the excessively long time required to extinguish the fire and to secure the space.

Numerous lessons can be learned from any casualty. Certainly in this case the OOD inport—and even the command duty officer—was not one of the more senior men on board ship. But what if he had been, as in fact one would expect the OOD inport to be during a weekend or on a holiday? Would he know what reports to expect from DC central? Would he know where and how to secure boilers topside? Or where and how to secure fuel oil to the boiler from the master quick closing valve topside? Or how to secure fuel to the emergency diesel or gas turbine generator with a QCV located outside the space containing each emergency generator?

General Specifications for Ships requires that boilers and fuel systems have valves that permit securing steam and fuel, respectively, from a remote location. Referring back to Figure 5–1, boiler stop valves which are designated as "A" valves allow the boiler to be secured from the damage control deck. Figure 3–2 shows how fuel to the boilers may be secured by tripping the master fuel oil quick closing valve. Emergency fuel oil QCVs permit securing both diesel and gas turbine emergency generators from a remote location. In addition, the air supply to a diesel generator can also be secured remotely because under certain conditions a diesel can sustain combustion by burning its own lubricating oil.

The inport OOD's involvement in engineering/propulsion plant light off consists of more than merely signing a "Man Aloft Request" and then watching the boiler technicians carefully climb the stacks to remove stack covers before lighting fires. The OOD should ensure that he is informed of the following:

- What boiler(s) is to be placed into operation.
- The time fires are lighted.
- The approximate time required to shift from shore power to ship's electrical power.
- The man who is in overall charge of the lighting off evolution, if other than the duty engineer officer.

In the case of auxiliary steaming, the OOD inport should ensure that he knows what equipment is in operation. In the majority of conventionally powered surface warships auxiliary steaming is similar to normal steaming in terms of fireroom equipment that must be operated to support the steam cycle. (One notable exception exists in the case of the FF-1040 and FFG-1 classes, which have an auxiliary boiler to furnish steam inport and a large diesel generator for ship's power.) The OOD should also know what generators are on the line. Auxiliary steaming also requires that an emergency ship service generator be available. The OOD should know the location of all major equipment necessary to support the auxiliary steaming condition. He should ensure that he knows how to secure fuel to the boiler and emergency generators, including air in the case of diesel generators, and the superheated and desuperheated steam stops of the on-line boiler, all from the proper remote actuating locations.

The threat of flooding a main engineering space exists during auxiliary steaming as well as during cold iron periods. From his knowledge of the main drainage system required by All Hands DC PQS, the OOD knows that proper lineup of the main drainage system will enable any main engineering space to be dewatered by using the eductor in another engineering space. The OOD should know the location of the remote-operated isolation valves in the main drainage system.

Certainly, the OOD inport has an inport emergency/repair party and other knowledgeable men on board who have the responsibility of controlling casualties that can occur during periods of cold iron, plant light off, or auxiliary steaming. Perhaps he will never have to use his knowledge of how to secure a boiler from a remote location or how to open valves in the main drainage system in order to dewater the forward fireroom of a DDG from the after engineroom. And then again, perhaps he will.

An inport OOD who understands something about the engineering plant in his ship is an asset to his ship. He's the kind of OOD who notices heavy black smoke belching from a stack during light off, hears a "panting" noise coming from the associated fireroom, and says to himself:

"Something's wrong. One of the BTs told me once that heavy black smoke can mean insufficient combustion air and perhaps the possibility of an explosion in the boiler air casing. My shipmates may be in trouble down there—I better call away the inport emergency party. . . ."

Rather than: "Those engineers are fouling up again! They better hope the commodore or the admiral doesn't see all that smoke!"

THE FIREMAIN SYSTEM

A basic knowledge of the firemain system is important to the OOD inport. The firemain performs the following functions:

1. It provides seawater for firefighting protection throughout the ship, including sprinkler systems in magazines and missile houses.

2. It operates in conjunction with eductors in drainage systems to

dewater both main and secondary engineering spaces, as well as other selected spaces where the potential for flooding exists.

3. It provides cooling water for equipment through a firemain/cooling main reducing station. In some ships this is the normal source for cooling water; in other ships cooling water from the firemain through a reducer is a back-up source if a salt water cooling pump failure occurs.

4. It provides flushing water throughout the ship. (*Spruance*- and *Perry*-class warships, however, use fresh water in the advanced waste disposal systems in these ships.)

5. Other important uses of the firemain may include the following in various classes of surface ships: emergency cooling for emergency diesel generators, back-up cooling for refrigeration plants, and the means of ballasting empty fuel oil tanks.

There are three basic types of firemain systems used in warships. Single main systems, which consist of a firemain that runs the length of the ship, are found in 1200 PSI CGs, DDG/DDs, and most smaller warships. A horizontal, loop-type system is employed in *Spruance*-class DDs. (This configuration consists of two single fore-and-aft firemains that are located in the same horizontal plane but are separated athwartships as much as possible; the two mains are cross-connected at various locations.) Larger ships usually have either a horizontal or a vertical loop firemain system, or a variation of the two.

In ships with either the single main or the horizontal, loop-type firemain system the firemains are located either on the damage control deck or a deck lower than the damage control deck. In ships with vertical loop firemain systems, the lower main is located below the lowest complete watertight deck, and the upper main is located below the highest complete watertight deck.

The diameter of the firemain may decrease in the forward and after sections of the ship in order to maintain adequate pressure in branches and risers. Fireplugs located throughout the ship are connected to the firemain through risers. Sections of the firemain that have been damaged or require maintenance can be isolated and bypassed by rigging a firehose jumper from fireplugs on each side of the area to be isolated.

Firemain pressures vary according to ship class. In 1200 PSI warships, firemain pressure is maintained at 100 psi. Normal firemain pressure on most larger warships is maintained at 125 psi. Firemain pressure on board *Spruance*-class DDs is maintained at 175–200 psi.

Pressure is provided by fire pumps located in various spaces throughout the ship. Isolation valves permit maximum sectionalization of the firemain for battle and damage control purposes. (Like many other systems, the firemain can be operated either split plant or cross-connected. Twin-plant warships normally have the firemain split during normal steaming under way or during inport periods by closing an isolation valve located in the forward engineroom.) Motor-driven fire pumps, which in many cases are the only type fire pumps installed on board the most modern warships, permit all classes of warships to maintain their own firemain pressure during periods of cold iron.

Ships such as 1200 PSI CGs and DDG/DDs have four motor-driven fire pumps. One electric fire pump is located in each engineroom; the remaining two are located in a pump room forward and aft, respectively. These ships also have a turbine-driven fire pump located in each fireroom for use when the ship is under way or steaming auxiliary inport.

An automatic start feature is provided on certain motor-driven fire pumps on surface missile ships to ensure that there will always be sufficient firemain pressure to combat accidental ignition of missiles in the magazines (missile booster suppression). These particular pumps are generally activated by a pressure switch when firemain pressure drops below a set point. Most electric fire pumps, however, have low voltage protection (LVP will be discussed in the following chapter); this means that the pump motor must be restarted manually even if the ship's electrical load is regained immediately after a momentary loss.

Electric fire pumps present some particular problems to the engineers, partially as a result of the ease with which one of these pumps can be put into operation.

At sea many engineers keep electric fire pumps on the line rather than using a turbine-driven fire pump (when installed) to share the load, since it is much easier to light off an electric pump than a turbine-driven pump.

When the ship returns to port and goes cold iron, the motor-driven pumps remain on the line—only now the pumps in use are located in the enginerooms where each can be monitored by the cold iron watch, rather than in the more remotely located pump rooms.

Exclusive use of a motor-driven fire pump located in the engineroom results in eventual failure of the pump. Then the pump must go to the tender for a motor rewind, or else the pump casing is so eroded that discharge pressure is affected. As luck would have it, that pump is also required to automatically augment firemain pressure for the missile booster suppression system.

As with other equipments, operation of fire pumps should be rotated regularly. This is good engineering practice, according to the Naval Sea Systems Command, which became so alarmed about the failure rate of motor-driven fire pumps in 1970 that increased fire pump reliability became one of NavSea's ten top projects.

Because of the extreme importance of the firemain system, the engineer officer and damage control assistant request divers to inspect fire pump sea chests and strainers at each tender or repair ship availability. If the ship has been inport for extended periods—or has been maneuvering in shallow water—marine growth or foreign matter may foul any salt water suction. In some cases marine growth and other debris can prevent check values in eductor systems and relief valves in the cooling main from reseating properly, as we have already seen.

A good understanding of his firemain system is a valuable asset for the OOD inport. The SIB contains much valuable information on this subject.

THE SHIP SERVICE ELECTRICAL PLANT INPORT

Imagine a routine working day inport: the time, 1015; the engineering plant status, cold iron.

Suddenly the ship goes dark. Alarms clang. The engineer officer, racing across the quarterdeck to the nearest below-decks access, charges into the fiery-eyed executive officer. The Captain, in service dress whites and headed for a change of command in another ship in the squadron, stalks up and asks tersely: "What the devil's going on?"

You are the OOD. The Captain's question echoes your sentiments exactly. But there's still further trouble when the duty cook appears and demands: "When do I get my ovens back on—don't you know 'Dinner for the Crew' is supposed to be piped down in less than an hour?"

Actually, any one of a number of events may have happened to cause that interruption in electrical power. But right now many men want that power restored, and the sooner the better. It helps if the OOD understands something about the ship's service electrical plant inport. His knowledge won't cause electrons to flow through the various equipments that comprise the electrical distribution system. However, the OOD's knowledge may be an important element of the teamwork required to get electrical service restored, or to prevent loss of electrical power in the first place.

A total loss of electrical power occurs only rarely when the ship is steaming auxiliary inport. Sufficient numbers of engineers are on watch to restore power or to deal with a casualty that results in loss of power. Moreover, the ship service diesel/emergency generators will be set up for automatic start when the ship is steaming auxiliary; so at least some vital electric service will be retained in the event of a casualty that forces a turbogenerator to be tripped off the line.

In general, loss of electrical power is more probable during periods of cold iron, when engineers other than those on board your ship are providing your ship with electricity. When a ship is receiving electrical power during cold iron, power from either the pier or the ship furnishing the electricity enters your ship through the *shore power terminal box*. This terminal box, which is located topside, connects to the *shore power bus*. The *shore power bus tie*, which is located inside one of the ship service switchboards (or switchgear groups), makes the power available to the ship service switchboard (or switchgear) for normal distribution.

When the ship is receiving shore power, the OOD should be aware that problems can arise. Shore power cables can be defective, and connection/disconnection of shore power is an exacting evolution that requires supervision by the leading electrician's mate. The amount of shore power available to a ship is generally limited. In addition, emergency electrical power from your ship service emergency generators will not (normally) be available, except in the newer classes of warships in which the diesel/emergency generator has a feature that permits the paralleling of that generator with shore power.

Reasons why problems can arise when the ship is receiving shore power are as follows:

1. The amount of shore power available to a ship is usually *limited.* If a total of 10,000 amperes (amps) is available at one pier, and only two cruisers and four destroyers are moored at that pier in a cold iron status, sufficient shore power is available for these six ships. But if two more cruisers and two more destroyers moor at that pier and shift to cold iron, the situation becomes critical.

2. The *limited amount* of shore power that the ship receives can cause problems. As an example, assume a landing ship dock (LSD) is capable of receiving a maximum of 1600 amps from the pier. Throughout the night the normal load is maintained at 700 amps. But in the morning, the gunner's mates begin transmission checks on their gunmounts. Cooks begin cooking. The massive stern gate is being lowered and raised to satisfy weekly PMS requirements. Other work centers are busy at their tasks, too.

Total load is now only about 1200 amps, according to switchboard meters. Sometimes, the amperage drawn increases to 1550 amps during load surges. But so far, so good. That's when the radiomen—the ship is maintaining its own radio guard—report rising temperatures in the radio equipment rooms. The engineering department's auxiliary gang (A-Gang) hurries to light off another air conditioning unit.

Suddenly lights go out, alarms are clanging, engineers running, and the Captain and cooks complaining. The probable reason: The starting current of many general-purpose alternating current motors on board ship is often four to six times higher than that required when the motor is running under normal load. Perhaps another work center energized more equipment just as an A-Gang man lighted off that air conditioning unit. In any event, men forgot to take those load surges into account. In this case, the OOD was unaware that a system of selective light off for various departments may be required during periods when only limited amounts of shore power are available.

3. The *shore power cables* which deliver power to the ship are ex-

tremely damage-prone. The electrician's mates are required to conduct periodic checks of the shore power cables. Chafing as the ship rides on her mooring lines can damage the cables. In other cases, cables that are carrying their maximum rated load may begin to overheat on extremely hot days. The OOD can prevent loss of shore power by noting abnormal conditions and informing the duty engineer of possible shore power cable failure.

4. *Connection/disconnection* of shore power involves significant hazards to both men and equipment. The OOD must ensure that the leading electrician's mate is on hand to supervise either evolution. Before the shift to shore power, terminal boxes must be tested to ensure that no voltage is present. Each cable must be properly insulated. Then the cable is energized to determine the phase and polarity of the shore power. The leading EM must supervise all phases of shore power connection and disconnection.

5. Finally, the OOD should be aware that the emergency ship service generator(s) installed in his ship may not have a design feature which permits automatic load paralleling with shore power when that generator is lined up for automatic start. If an emergency generator has this feature, it should be lined up for *automatic start* even when the ship is "cold iron" in order to ensure that a source of power will be available to vital auxiliaries such as fire pumps should there be a loss of shore power. Otherwise, the emergency generator must be maintained in the *locked-out* status.

The 1200 PSI FF/FFGs have ship service diesel generators which should be set for *automatic start* when the ship is cold iron. But in the majority of other conventionally powered surface warships, maintaining the emergency generators in *lock out* while receiving shore power is required to protect shipboard equipment. If the emergency generator were incorrectly set for automatic starting while receiving shore power, a temporary loss of shore power would cause the generator to light off. Resumption of shore power could then result in damage to shipboard equipment because of phase or polarity differences in the two sources of power.

In extreme cases, the incoming shore power could drive the emergency generator as a motor. Electrician's mates call this "motorizing a generator." Unless its KW output is not exceeded, "motorizing" the generator will not damage the generator itself. However, the prime mover—either a diesel or a gas turbine, depending upon the emergency generator installation—and the transmission that drives the generator end would be damaged.

(Emergency generators can be used to augment shore power, as is sometimes done on board ship to test the emergency generator under a load and to satisfy additional power requirements when sufficient shore power is not available. However, the electrician's mates must ensure that the selected load is isolated from shore power before the emergency generator is used to *feedback* to the main switchboard from its own switchboard to energize the selected load.)

SECURITY OF ENGINEERING SPACES

The engineering plant is vulnerable to persons who seek to prevent a warship from performing its mobility missions. The OOD inport should realize that the plant is more vulnerable to willful destruction and damage inport than under way. The plant is especially vulnerable during periods of cold iron.

The Naval Sea Systems Command and type commanders have published broad guidelines concerning locked spaces. As a general rule, steering gear rooms and shaft alleys are locked inport. Emergency generator rooms and gyro/IC rooms may be locked when they are unmanned or equipment is not in use, depending on a particular ship's policy. Even secondary access to a main engineering space may be locked, if the cold iron watch can keep only the main access under general surveillance.

Within a space itself, certain equipments are always kept locked unless opened for either inspection or maintenance. The high security padlocks on the inspection ports of the main reduction gears are the most familiar example of engineering plant security. There are, however, numerous other equipments and valves in certain systems that are locked as required by ship's policy and higher authority.

Access to engineering spaces should generally be limited only to the men who work in those spaces. Men working inside an engineering space that is secured must *always* be able to leave the space without using a key, and that space may *never* be locked from the outside when men are inside. A damage control party must *always* be able to enter a space by using a key or damage control tools to force a lock on the outside of an unmanned space.

Flooding alarms and rate of temperature rise or smoke detectors, where installed, enhance security. But cold iron watches, sounding and security watches, and random tours of engineering spaces by senior supervisors will always be necessary to maintain security of engineering spaces.

The fact that security of the engineering plant must be emphasized is not a pleasant subject. A warship that misses a commitment because of willful damage to its propulsion plant is not a pleasant subject either.

FUEL HANDLING

Many commanding officers prefer to refuel their ships at sea rather than inport.

The reasons are numerous: Underway replenishment is an exacting operation, and a ship retains proficiency through practice. During an UNREP the fueling "first team" is on board; because of this the danger of fuel spillage is reduced. Moreover, accidental spillage of *small* quantities of fuel at sea does not pose the same threat to the environment that a similar *small* amount spilled inport would.

However, reduction in the number of replenishment ships and underway time for all ships has decreased the opportunities available to refuel at sea. Instructions issued by fleet commanders require that ships inport (and at sea) maintain a certain percentage of fuel on board at all times. As a result, the OOD inport may have a refueling scheduled during his watch.

The engineering department maintains fueling checklists. If an inport fueling operation is scheduled, which could be only an internal transfer of fuel oil, the OOD should consult the applicable ship's instructions to review his responsibilities before the operation begins.

The OOD's involvement in a fueling operation can entail more than merely hoisting the BRAVO flag and passing word concerning the "smoking lamp" at periodic intervals. Time is critical when dealing with even the smallest oil spill. If fuel is spilled, firehoses should be used to contain the spill between ships in the nest, a quay wall, or a floating object such as a camel or boom. The OOD may also be required to launch a boat quickly so the pollution control party can apply the proper collection and absorption materials and techniques.

RESPONSE TIME

Even though the OOD inport may be assigned to a department other than the engineering department, he should have a general idea of the time required for the engineers to light off their propulsion plant in order to get under way.

Ships moored inport are required to be able to get under way within certain time frames. If a ship is alongside a tender and cold iron, for example, it may be on a 48-, or even 96-hour, steaming notice. If a state of national alert or other emergency is declared, the senior officer present afloat (SOPA) may direct that even ships undergoing a tender availability get their plant ready to light off and be prepared to get under way on short notice.

Weather is certainly a factor, too, on the response time required for the engineers to be able to get under way. When heavy weather is imminent, the OOD inport should keep himself advised of the time required to get under way on the minimum boiler (propulsion) power required by the commanding officer and SOPA. The following example illustrates why:

Knight's Modern Seamanship, in a discussion of wind force or loading, states that for practical purposes wind force can be expressed by the following equation:

$$F=AV^2$$

where A equals the projected area above the waterline in square feet, and V equals the wind velocity in knots. Since A, the area, tends to remain constant for a ship riding at a buoy, anchored, or moored alongside, wind velocity is the important variable.

If a ship is moored to a buoy in 5 knots of wind, no problem exists. If the wind doubles in strength to 10 knots, the ship still rides comfortably, even though the strain on the anchor chain is now *four* times greater than the strain at 5 knots. But suppose the wind increases to 25 knots, a fivefold increase over its original velocity of knots: the strain on the anchor chain is now *25* times greater than the strain at 5 knots, and the situation begins to get tense.

General Specifications for Ships require that anchor chains be strong enough to hold in 70 knots of wind. But by the time the wind has increased to 25 knots, the Captain or the command duty officer probably will have directed that a propulsion plant be placed on the line. Perhaps the Captain will decide to "steam to the buoy," using his engine(s) as a means to reduce strain on the anchor chain if the wind increases in velocity. Perhaps the Captain will get under way while it is prudent to do so, while his engine(s) have a reserve of power and his maneuvering is relatively unrestricted.

Light off of any propulsion plant takes time. In the case of a conventional propulsion plant, normal light off from cold iron using EOSS takes about four hours. In an emergency, the engineers can have the main engine(s) warmed up and the screw(s) ready to turn in an hour. If the ship has been steaming auxiliary, a destroyer or similar ship can be ready to get under way cross-connected on one boiler in about 40 minutes.

Engineers in FF-1040 and FFG-1 class warships are justifiably proud of the responsiveness of their steam generators. These ships can have full boiler power available from cold iron status in about 45 minutes, using emergency EOSS lighting-off procedures.

The most responsive propulsion plants are the gas turbine and diesel-powered plants. Engineers in the *Spruance*- and *Perry*-class warships can have full propulsion power available from a cold iron status in less than 15 minutes.

Inport or under way, the OOD knows weather conditions have a bearing on the desired degree of engineering readiness. The OOD inport should have a general idea of engineering plant status, so he will know how long will be required to light off a propulsion plant if heavy weather is imminent.

CHAPTER FOOTNOTE

This chapter has been an "inport" chapter, one in which topics such as auxiliary steaming, the importance of the firemain, and cold iron were introduced. The intent of this chapter was to acquaint the OOD inport with some of the information about his engineering plant required to keep him out of "hot water," even during periods of cold iron.

The following chapter is an "underway" chapter, one that contains topics such as basic electrical distribution, water production, and steering. Electricity is vital on board ship—for combat systems, propulsion, water production, steering, even for baking bread and rolls.

In order to get under way again, permit me to recount an experience with a ship's baker a number of years ago. Imagine that he was a shipmate of the distraught cook who confronted you on the quarterdeck just after your ship had lost the electrical load inport, in a previous section of this chapter:

I was officer of the deck of a destroyer assigned primary rescue station for a CV conducting nighttime flight operations. The night was clear, but the wind so strong and the seas were so rough that the carrier only needed to make about 10 knots while launching aircraft.

Maintaining station was difficult enough. When the carrier turned upwind or downwind, our ship and the destroyers in the ASW screen really began to bounce. Down below, no one was getting much sleep.

Suddenly a figure in white loomed beside me. It wasn't the Captain. But there was plenty of authority in the voice:

"You the OOD?"

"Yes," I replied. "Anything I can do for you?"

The answer came quickly:

"One thing—just take it easy. I've already had coffee cake spill over in the oven three times because of the way you've been driving this ship.

"If it happens just once more, the crew ain't gonna get fed. And I'm going to tell the Captain it's your fault!"

Then having made certain that I knew the "heat" was really on me, our ship's baker returned to his ovens.

CHAPTER REFERENCES

Knight's Modern Seamanship, Fourth Edition
Principles of Naval Engineering, NavPers 10788B
Shipboard Electrical Systems, NavPers 10864C

7

Concerning Electrons and Molecules

The propulsion plants of modern warships are more dependent upon motor-driven auxiliary equipment than were conventional plants of the past. In the most modern combatants, such as the *Spruance-* and *Perry-*class warships, the use of electrical and electronic controls as well as motor-driven auxiliaries for propulsion makes the propulsion plants of these ships even more dependent upon electrical power.

Even if electrical power were not so vital to his propulsion plant, the officer of the deck of a warship would still require a basic understanding of the importance of his electrical plant to his ship's combat capability, maneuverability, habitability, and survivability. As an example, imagine a warship without electrical power in a hostile action, and you suddenly visualize the grim picture of a burning, foundering hulk:

In the dark and silent combat information center, electronic warfare specialists would find their electronic intercept receivers useless in detecting illumination of their ship by enemy fire control radar. Lookouts might detect the fireball of a cruise missile as it plunged toward their ship, but detection by other sensors would be impossible. Modern weapon systems could not react to the threat. The ship could not maneuver. Electric firepumps could not provide firemain pressure to fight fires or to control flooding. And finally, the ship would become a sinking, rather than foundering, hulk.

Of course, such a calamity could be caused only by catastrophic failure of the ship's entire electrical generating capability and distribution systems. Like the propulsion plant, the ship's electrical plant features equipment duplication; physical separation of equipments; and "plant insurance policies," such as emergency generating and distribution systems and safety devices to protect equipments. Like the propulsion plant, the electrical plant of a modern warship is a highly reliable, responsive, and durable installation.

This chapter will present, in general terms, information about the electrical plant of interest to the OOD or the prospective OOD. Topics concerning the electrical plant include the following:

Electrical generation
Electrical distribution
Split and parallel operation of the electrical plant
400-Hz power

Electricity is also vital to maneuvering and to water production on board ship. The majority of steering systems in the U. S. Navy are of the electrohydraulic variety, with variations of this basic type. The production of boiler feedwater and potable (fresh) water from seawater occurs in distilling units, which employ pumps to perform various important functions in the distillation process. Because of the dependence of both steering and water production on electricity and the importance of these subjects to the OOD, the following additional topics are also presented in this chapter:

Electrohydraulic steering gear
Water production
Water management

This chapter, then, contains information "concerning electrons and molecules" of interest to the OOD. Information presented in this chapter will enable the OOD to see the relationship of the electrical plant to the propulsion plant, as well as the importance of the electrical plant to weapons systems, steering, and water production. Additional information on specific equipments can be obtained from the Ship's Information Book (SIB) and technical manuals for equipments on board specific classes of ships.

The OOD's understanding of his electrical plant begins with a basic understanding of electrical generation on board his ship.

ELECTRICAL GENERATION

Most electric power used on board ship (and ashore) is generated by alternating current (AC) generators. With very few exceptions, these generators are 450-volt, 3-phase, 60-hertz (Hz) installations.

Regardless of whether the chief electrician refers to the equipment as an alternator or simply as an AC generator, the principle of operation is the same: a magnetic field cutting through conductors, or conductors passing through a magnetic field. All generators have at least two distinct sets of conductors. In one set, called the field windings, direct current is passed to create an electromagnetic field of a fixed direction. Output voltage is generated in the second set of conductors, referred to as the armature windings.

The relative motion that must exist between the armature and the field in an AC generator is provided by two major assemblies. The *stator* is a nonmoving armature. The *rotor*, rotating inside the stator, is the rotating member in which the field is applied. The rotor is driven by the prime mover, which may be either a steam turbine (SSTG), gas turbine (SSGTG), or diesel engine (SSDG).

The revolving-field AC generator is the most common ship service generator. In this type of ship service generator, direct current from a separate source is passed through windings on the rotor by means of sliprings and brushes (which do not resemble conventional brushes, but rather are tightly bonded and formed pieces of carbon). This creates a rotating electromagnetic field, which cuts through the armature windings embedded in the stator. Alternating voltages induced in the fixed armature windings represent output power that can be removed by means of fixed *terminals*.

The major advantage of a revolving-field AC generator is that large ouput voltages can be obtained from a relatively small piece of equipment. Another important advantage is that output power is removed by fixed terminals rather than by sliding contacts. This makes the output circuit easier to insulate, thereby minimizing the danger of fire from electrical arc-over.

The output from the ship service generator terminals is connected by a generator circuit breaker to the generator's associated switchboard. *Bus ties* allow one switchboard or switchgear group to feed power to one or more other switchboards. Bus ties also can be used to connect two or more switchboards or switchgear groups so that their associated generators can supplement each other electrically. (This is called "parallel" operation.)

Before we move on to electrical distribution, some additional points concerning generators may be helpful to officers of the deck:

• The most efficient way to regulate the voltage output of an AC generator is to control the strength of the rotating magnetic field. A relatively small amount of DC voltage and current used to vary the field can control a large AC voltage. Amplidynes and combined static excitation/voltage regulation installations are two types of automatic voltage regulators widely used on board naval ships. (The combined static excitation/voltage regulation installations are used on board the more modern warships.)

• Most naval ships have at least two different types of prime mover for ship service generators. Even though usually considered as emergency generators in carriers, cruisers, and conventionally powered DD/DDG/FF/FFGs, the diesel (and gas turbine, where installed) generators may be used for ship service power. (In *Spruance*-class DDs and *Perry*-class FFGs gas turbine–powered generators [SSGTG] and diesel-powered generators [SSDG], respectively, are the only ship service generators.)

• Generator prime movers require protection. In the case of turbogenerators a high water casualty poses the same threat to the SSTG that it does to the main engine. Turbogenerators must also be protected by *back pressure trips* in case vacuum is lost in their associated condensers. *Overspeed trips* protect generators by securing the *prime mover of any type of generator if it overspeeds.*

• When generators are operated in parallel, *reverse power relays* trip individual generator circuit breakers to prevent one generator from motorizing another. This actually protects the prime mover and associated reduction gearing rather than the generator itself. (Note: Turbogenerators [SSTG] and generators powered by another type of prime mover should *not* be placed in parallel operation with each other except in the case of the 1200 PSI FF/FFG and some other ships where this design feature has been incorporated.)

• In 1200 PSI ships the capacity of the ship service generators ranges from 500 KW SSTGs in 931-class DDs to the 2500 KW SSTGs on board CVs. The majority of 1200 PSI CGs have a total of four 1500 KW SSTGs for primary generating capability with two smaller generators powered by different prime movers for emergency use. *Spruance*-class DDs have three 2000 KW SSGTGs with all three capable of parallel or split plant operation.

ELECTRICAL DISTRIBUTION

Chapter 2 described the emergency electrical generation and distribution system as a plant "insurance policy." The simplified illustration of Figure 7-1 (used in that chapter as Figure 2-4) shows the relationship between the ship service (SS) and emergency power system (EPS) in conventionally powered DD/DDG and 1200 PSI CGs.

The power distribution system connects the generators (SSTG, SSGTG, or SSDG, depending upon the prime mover) that supply power with the equipments that use electrical power. Power distribution systems on board naval ships are actually composed of the ship service distribution system, the emergency power distribution system (described in Chapter 2) and the casualty power (CPS) distribution system.

The 120-volt lighting distribution system is supplied from the various power circuits through banks of step-down *transformers.*

Engineer officers of warships can sum up their electrical distribution system in a short phrase: duplication designed to enhance reliability. These chief engineers point out that:

• Switchboards and associated generators are located in separate spaces

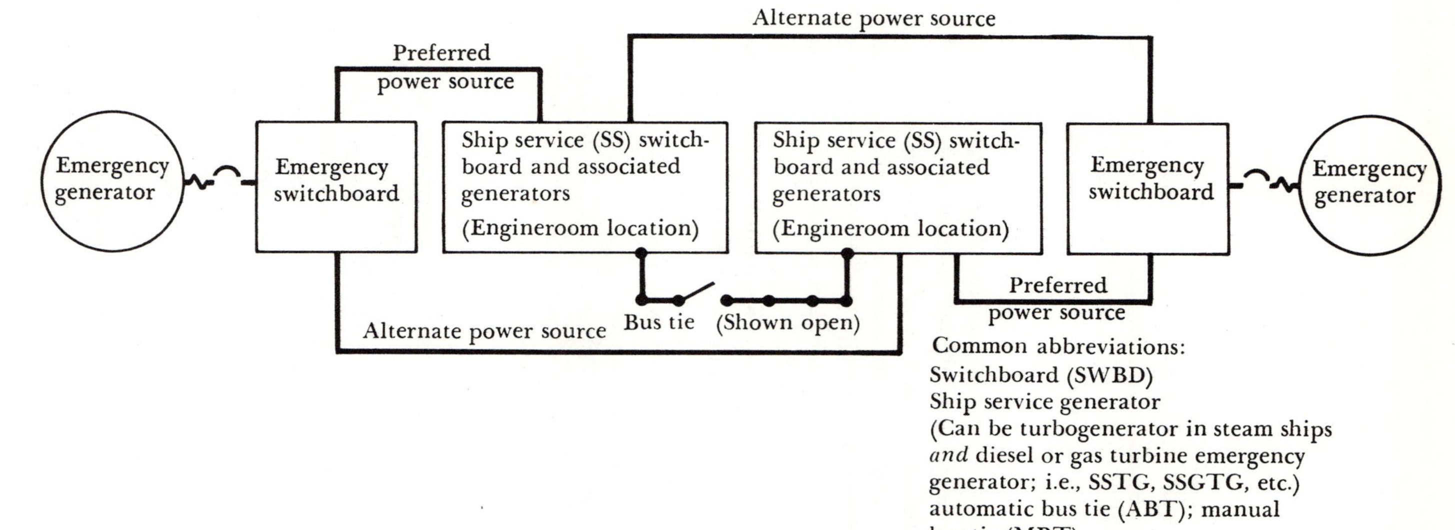

Figure 7–1. Simplified 1200 PSI DD/DDG power distribution

to minimize the possibility of an enemy hit or another casualty knocking out the entire electrical plant.

• *Normal* and *alternate* SS *feeder* cables located below the waterline conduct the power along opposite sides of the ship. In addition, an *emergency* feeder cable is located near the centerline and well above the waterline in many warships.

• Ship service generators and distribution switchboards are interconnected by bus ties, so that any switchboard can be connected to feed power from its associated generator to another switchboard.

• Generators designated as "emergency generators" can provide power to a main SS switchboard in warships by means of a manually operated "feedback" tie. The limited capacity of an emergency generator allows only various selected *vital loads* to be energized during these periods, however.

• At least two independent sources of power are made available to the various pumps and other equipments designated as vital loads. Power may be available from either a *normal* and *alternate,* or *normal* and *emergency,* or a *normal* and *alternate and emergency* feeder, depending on the load.

Because of these features, electrical plants of warships are highly reliable and dependable.

In newer and larger ships various equipment and systems are energized from *load centers,* rather than directly from SS switchboards or switchgear groups. This arrangement requires less weight, space, and equipment than would be required if each load were connected directly to a main switchboard or switchgear group.

Regardless of the type of ship, bus transfer equipment is needed at the load center, distribution box or panel, or the load itself to select a source of power. Automatic bus transfers (ABTs) detect voltage drops and automatically switch to the stronger of two sources. Manual bus transfers (MBTs) must be operated by men on watch.

Both manual and automatic bus transfer equipment is used in the steering installations on board most warships. Motors for pumps used in the steering installation can receive electrical power from either normal (also called "preferred") or alternate or emergency sources. MBTs are used to select between the normal and alternate power sources for steering motors (and other motors). If, for example, engineering doctrine on board a particular DDG required that power to steering motors be made available from the forward ship service switchgear group rather than the after switchgear group, the normal source of power for steering motors, an electrician's mate would position the MBT for steering equipment to the alternate position. Electrical power for the steering motors would now come from the forward ship service switchgear group. If that ship service power source were interrupted or lost, an ABT would automatically transfer the steering to that equipment to ship service power from an emergency source.

Protection for motors in the steering system, as well as for motors for other vital equipment throughout the ship, is provided by either low voltage protection (LVP) or low voltage release (LVR). LVP secures a motor when voltage to it drops below a certain value. That motor must then be restarted manually. LVR secures a motor also, but a motor equipped with LVR will restart automatically when voltage is restored to it normal (or a predetermined) value.

Officers of the deck need not be too concerned with LVP or LVR, although the OOD should know that the steering gear motor of the unit in use is protected by LVR. This design feature incorporated in steering systems is an additional feature that helps to ensure the reliability of the steering system.

Other key members of the ship's mobility team, however, are concerned with LVR and LVP, as well as with ABTs and MBTs. For example, propulsion plant supervisors should know which of their motor-driven pumps have LVP and which have LVR. Space top watches should also know which equipments receive electrical power via ABTs or MBTs, as well as the location of all bus transfer equipments within their space. The OOD's propulsion plant partner, the EOOW, must know what vital and semivital equipment is energized from a particular power source. The EOOW must also ensure that the electrical plant is configured for either split plant or parallel operation, depending upon the direction of the engineer officer.

The OOD needs a general idea of his electrical plant's capabilities. He should know the total KW capacity of the generators, as well as how much load is placed on the electrical plant during Condition III and under battle conditions.

With regard to electrical distribution, the OOD should know the location of power sources available to his weapons systems and steering and whether alternate power is available from either an ABT or an MBT. The OOD should know whether his electrical distribution system is being operated in either a split plant or parallel configuration.

SPLIT AND PARALLEL OPERATION OF THE ELECTRICAL PLANT

All OODs understand the terms "split plant" and "cross-connected" when used with reference to lineup of the main and auxiliary steam systems. The terms "split plant" and "parallel(ed)" are used to describe the configuration of the electrical plant in much the same manner:

Imagine that a bus tie between switchboards functions rather like cross-connect piping in a main steam system. The electrician's mate of the watch (EMOW) uses his bus tie circuit breaker in much the same manner that fireroom or engineroom watchstanders use a cross-connect valve (some engineers prefer to say "split plant valve") in a main steam system. The major difference in the analogy is this:

By *closing* the bus tie circuit breaker the EMOW connects the switchboards electrically, allowing the entire load to be *paralleled* between equally loaded generators.

By *opening* the bus tie circuit breaker the EMOW disconnects the switchboards from each other, dividing or *splitting* the load between generators.

There are advantages and disadvantages with both modes of operation. For example, in a 1200 PSI CG one electrician's mate can control the entire electrical load of the ship from the forward switchgear group (only) if the load is paralleled between a generator(s) in the forward plant and a generator(s) in the after plant. If the engineer officer specified that the electrical plant must be in a split configuration, a greater degree of reliability will be achieved for this class ship, but another electrician's mate will be required to stand watch on the after switchgear group.

In those warship classes in which total generator capacity exceeds normal steaming loads by a wide safety margin, use of generators in parallel (also called "cross plant operation") allows routine maintenance to be performed on an idle generator under way. But when total generator capacity does not provide such a safety margin, a boiler casualty or electrical casualty which required that SSTGs be tripped off the line could result in the loss of ship service power. (In such a case an emergency ship service generator would begin operation automatically, but power would be available only to selected vital equipments.)

If the electrical plant is operated in split plant configuration, a boiler casualty or switchboard casualty in one plant will not affect equipments fed from a switchboard/switchgear group in another plant. Moreover, ABTs for equipments in the plant in which the casualty occurred will shift to make an alternate source of power available to those equipments. In general, the ship's electrical load should be split for all increased conditions of battle readiness, for entering or leaving port, and for other evolutions that require an exacting degree of seamanship, such as underway replenishments or heavy weather operations.

Still, different doctrines and conditions exist for different types of ship. Ships with SSTGs operate with the electrical load split during conditions of increased readiness for the reasons given in preceding paragraphs. A greater degree of sophistication exists in the electrical plants of the *Spruance*- and *Perry*-class warships. The SSGTGs and SSDGs in these respective ship classes operate independently of the propulsion gas turbines. Advanced electronic controls and the ability of both gas turbine and diesel prime movers to accelerate rapidly from dead stop permit parallel operation of the electrical plant in the Navy's gas turbine–powered warships under conditions that would not be acceptable on board conventionally powered DD/DDGs, CGs, and FFGs.

400-HZ POWER

Although this book is concerned primarily with propulsion, it is written for the OOD or prospective OOD of a warship. Without its 400-hertz (Hz) power, a warship is little more than a helpless giant.

Prior to guided missile system installations, the only uses of 400-Hz power on board ships other than aircraft carriers were for the gyrocompass, degaussing, and synchro systems. Now, weapons systems, sensors, interior communications (IC) systems, and other electronic systems on board *all* types of warship are designed to operate on 400-Hz power. Aircraft servicing systems are also 400-Hz systems.

Electrician's mates refer to 400-Hz power after conversion by a motor generator, static inverter, or static converter, as a Type III power system, since it is a system designed to maintain stringent voltage and frequency tolerances. The present generation of fire control radars and other sensors, computers, and other precision equipment simply cannot withstand any voltage or frequency transient much greater than about 1% (400 ± 4 Hz). Motor generator sets convert the ship service 450-volt, 3-phase, 60-Hz power to 400-Hz power in the majority of warships in the U. S. Navy. Static inverters and static converters which utilize solid state components are replacing motor generator sets in the more modern warships.

(Static converters are used exclusively on board *Spruance*-class destroyers, converting the ship service 450-volt, 3-phase, 60-Hz power to 450-volt, 3-phase, 400-Hz power. Like static inverters, static converters represent an improvement over motor generators because of their lighter weight, [total] smaller space requirements, reduced maintenance requirements, higher efficiency, and faster reaction to transient disturbances. The use of solid state static coverters rather than motor generators with moving parts also reduces total ship self-noise, a major consideration for an ASW ship.)

On board ship the conversion of 60-Hz power to 400-Hz power occurs in either motor generator (M/G) rooms or load centers, depending on the class of ship. The 400-Hz power is distributed to various equipments by means of special frequency (SF) switchboards. Figure 7-2 illustrates basic 400-Hz distribution on board a CG 16 class cruiser.

In Figure 7-2, 60-Hz ship service power is available to M/G sets 1, 9, and 13 from ship service switchgear sections as indicated. M/G sets 1, 9, and 13, which are rated at 200 KW each, convert 60-Hz ship service power to 400-Hz ship service power and feed ship service special frequency switchboards 1SF, 5SF, and 6SF, respectively.

The CG 16 class cruiser also has seven 60-KW, 450-volt, 3-phase, 400-Hz

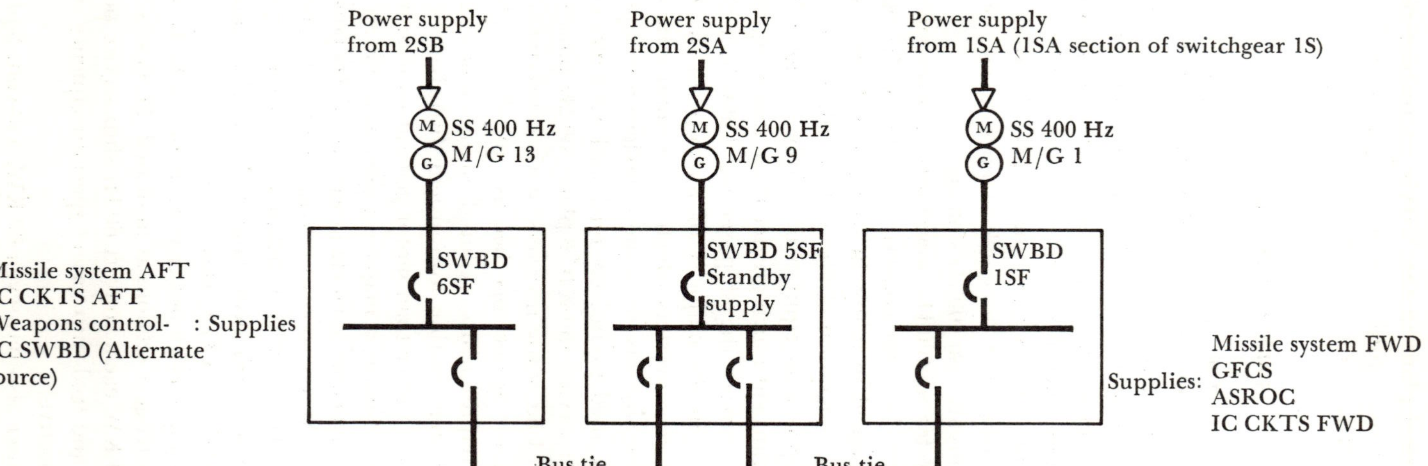

Figure 7-2. Sample ship service switchgear/motor generator set/special frequency switchboard bus tie connections (for CG 16 class cruiser)

motor generator sets, which are used as follows: M/Gs 5, 7, and 8 (as a standby) supply missile illuminator power switchboard 2SF for the forward missile battery; M/Gs 10, 11, and 12 (as a standby) supply missile illuminator power switchboard 4SF for the after missile battery; and M/G 14 is the normal supply for 3SF, the weapons control IC switchboard. If M/G 14 fails, an ABT transfers 3SF to its alternate power supply, switchboard 6SF.

Regardless of the ship class, duplication of 60 Hz–400 Hz converting units and physical separation of the special frequency switchboards provides the 400-Hz system with a large measure of reliability. Nevertheless, weaponeers and electronics personnel should be alert to trip their equipment off the line if the ship suffers a loss or major reduction in 60-Hz ship service power, since this will affect 400-Hz power.

ELECTROHYDRAULIC STEERING GEAR

Throughout this chapter the terms "vital load" and "vital service" have been used frequently. A ship's steering system is one of the most vital of all systems. Electrohydraulic steering systems, the type of steering system used on board the majority of naval ships, depend upon uninterrupted 60-Hz ship service power to perform this vital function.

Electrohydraulic steering gears in use include various types of equipment. Some shipboard installations have double hydraulic rams and cylinders such as that shown in Figure 7-3. Other installations are single-ram arrangements. Regardless of the type of installation, the principles of operation are basically the same.

In an electrohydraulic steering system the *steering* (running) *motor* turns at a constant speed. Assume that the *tilting box* within the *pump* driven by the steering motor is in a neutral position, and therefore there is no flow of hydraulic fluid.

When the helmsman turns his wheel, an electric signal is transmitted to the *synchronous receiver* in the steering gear room by the *port cable* or the *starboard cable*, depending on which was selected in the pilothouse. Movement of the synchronous receiver is transmitted to the *control shaft* by means of gearing in the *differential gear train*. Movement of the control shaft operates the tilting box within the pump. Hydraulic fluid is now forced into one pair of *cylinders*, and an equal amount of fluid is returned to the pump from the opposite pair of cylinders.

When the steering wheel and synchronous receiver stop moving, movement of the *ram* operates the *rack and pinion*. Gearing in the differential gear train transmits this movement of the rack and pinion to the control shaft. Movement of the control shaft now returns the tilting boxes to a neutral position, and the flow of hydraulic fluid stops.

In the installation shown in Figure 7-3, the rams act as hydraulic

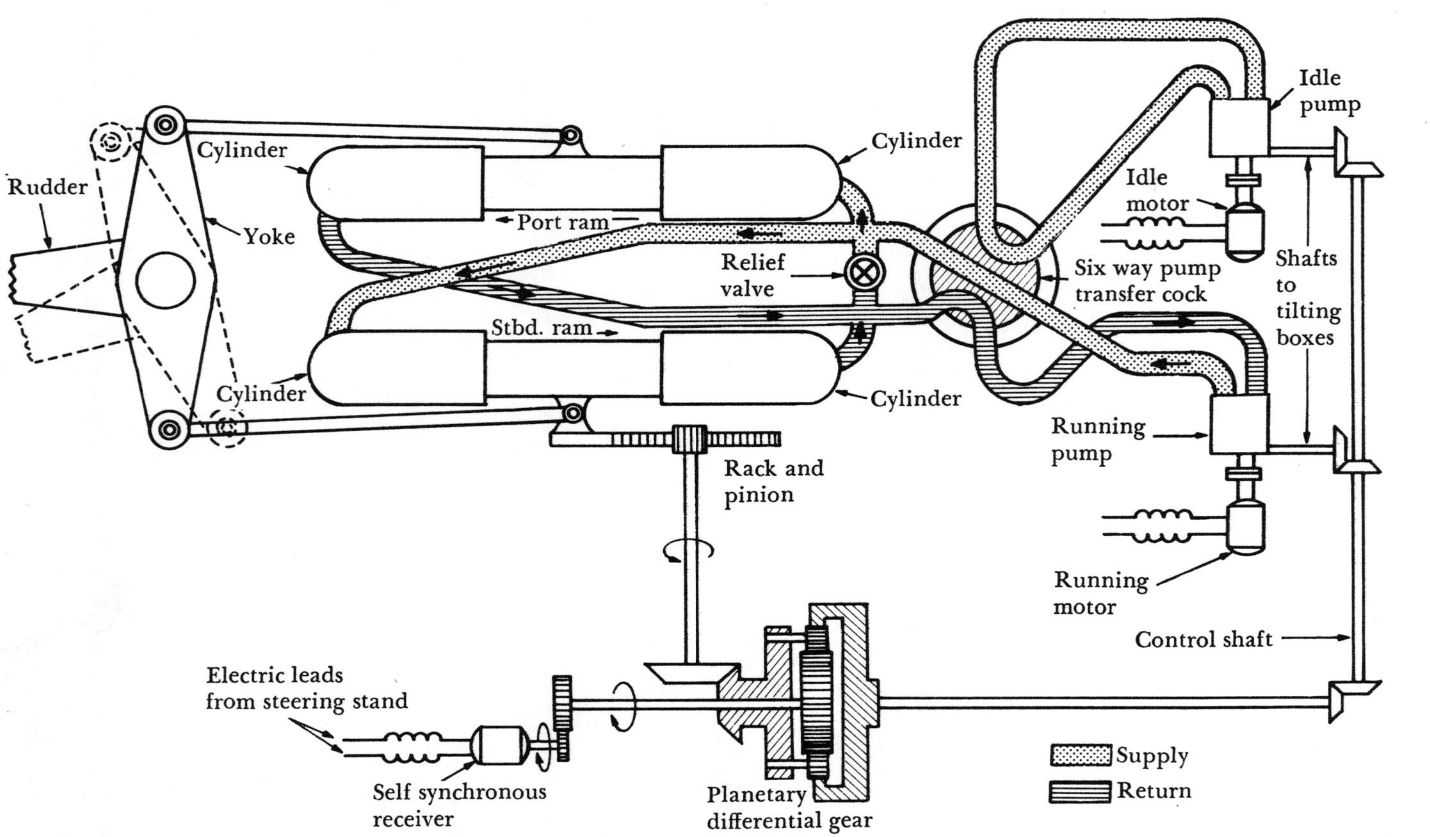

Figure 7–3. Simplified double-ram electrohydraulic steering gear (Illustration shows rudder being positioned to turn ship to starboard)

plungers. The rams are connected to the *yoke,* as is the *rudder.* A double-acting *relief valve* relieves the piping of any excessive pressure which might be caused by wave action on the rudder or by a jammed rudder. The *six-way transfer cock* enables the after steering watch to transfer quickly from the operating pump to the standby pump.

The after steering watch can also take control from his station by operating the *trick wheel* (not shown), which is geared directly to the synchronous receiver.

If electrical power is completely lost, steering is possible by using a *hand crank* to position the tilting box. (However, steering by use of the hand crank is a physically exhausting evolution, even at slow speeds in calm seas.)

As mentioned earlier in this chapter, steering equipment in most warships has normal, alternate, and emergency power supplies. MBTs are used to select which switchboard or switchgear groups will be the respective normal (preferred) and alternate ship service supplies. An ABT automatically transfers steering motors to an emergency ship service power supply if the normal (and alternate) power is interrupted. In most steering installations the motor in use features low voltage release; the standby motor has low voltage protection.

Electrohydraulic steering control may also be maintained with power available from the casualty power system (CPS). Connection of steering to a source of casualty power would only occur during General Quarters, however, as a result of battle damage to both the ship service and emergency power systems. Under such conditions the OOD would be concerned with numerous other considerations besides power to his steering gear.

Nevertheless, every OOD should have a thorough knowledge of the procedures for shifting steering control in his ship, as well as a basic understanding of how the steering gear operates. In most ships steering motors and transmission (steering) cables are shifted daily, usually at a time specified in standing orders. In addition, loss of steering control drills give watch sections the training they need to deal with actual casualties.

As a final note on steering, the OOD must control all drills involving loss of steering and the periodic shifting of steering units. A senior officer who since has had three commands at sea recounts this particularly harrowing experience, which happened when he was officer of the deck of a destroyer:

"We had been operating within visual signalling distance of the aircraft carrier throughout the day, conducting individual ship exercises such as Man Overboard drills and steering from After Steering.

"At twilight we received a signal to take station astern of the carrier. I took the conn from my JOOD, since our initial course to station was nearly the reciprocal of the carrier's, and high relative speeds were involved.

"What should have been a fairly simple maneuver became a tight situation. Rudder response seemed extremely slow. I ordered the Boatswain's Mate of the Watch to take the helm. Then I needed to make a hard turn and a flank bell to open the range to the carrier.

"That's when the Boatswain's Mate of the Watch shouted that After Steering still had control—my JOOD had passed steering control to After Steering during the drills, and we had not returned steering control to the Bridge when we received the signal to take station on the carrier!"

WATER PRODUCTION

Production of those molecules of water so important to the conventional engineering plant and to shipboard habitability depends to a great degree on the electrical plant. All shipboard distilling plants heat seawater (also called "feed") to the boiling point and then condense the resulting vapor to obtain distillate. Figure 7-4 is a simplified illustration of this process.

There are two general types of shipboard distilling plant. *Vapor compression* units, which have application in smaller ships where daily requirements do not exceed 4000 gallons per day (gpd), are operated by electrical energy. *Low pressure steam distilling* units, which are used in the great majority of both surface and subsurface warships, use steam at very low pressure as the energy source to produce evaporation. Low pressure steam distilling plants depend on motor-driven pumps to support the

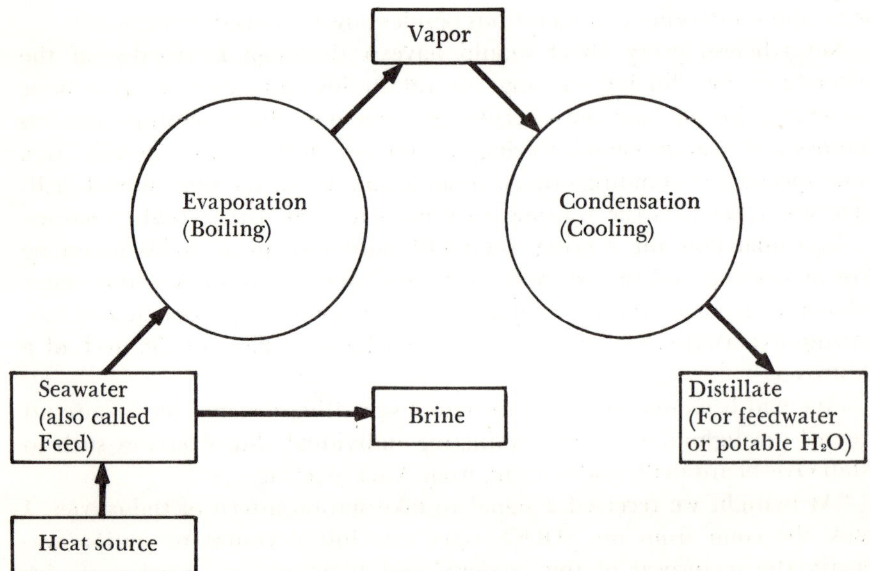

Figure 7–4. Simplified diagram of shipboard distillation process

evaporation-followed-by-condensation process that converts seawater to water for use by the engineering plant and potable (fresh) water.

Steam distilling plants include three basic types: submerged tube units, flash-type units, and vertical basket units. Each type consists of a heating section to produce evaporation, a condenser to condense the resulting vapor, a preheater to preheat the seawater (feed), and a cooler to cool the distillate.

All steam distilling plants also utilize air ejectors to establish and then sustain vacuum in the various stages. The air ejectors operate on steam that has been reduced to 135 psi (in most installations). The heating source, however, is steam at a much lower pressure. Auxiliary exhaust reduced to 5 psi is the primary heat source of distilling units on board conventionally powered surface warships.

Flash-type distilling plants, which have become the standard installation in all larger surface ships, have four motor-driven pumps: the feedwater pump, salt water heater drain pump, distillate pump, and brine overboard pump. The capacity of flash-type units varies from five-stage units producing 50,000 gallons each day in carriers to two-stage units producing 12,000 gallons daily in destroyers and many cruisers. Each class of ship may have multiple units.

In flash-type units feed (seawater or river water) picks up heat while passing through the tubes of the distillate cooler, the distilling condensers, and the air ejector condenser. Final heating is done by auxiliary exhaust admitted to the shell of the feedwater heater. From this heater the feed enters the first stage feed box and emerges through orifices in a flash chamber. As it enters, a portion of the heated feedwater "flashes" or vaporizes because the pressure in the flash chamber is lower than the saturation pressure corresponding to the temperature of the hot feed (that familiar temperature-pressure relationship, once again!). The vapor condenses on the tubes of the first-stage distilling condenser. Feed that does not vaporize then passes into the second flash chamber, where the pressure is even lower. The process is repeated in successive stages until the brine remaining in the final stage is removed by the brine overboard pump. Vapor formed in each stage passes through a vapor separator into a stage condenser where it becomes distillate, The distillate is removed from the final stage by the distillate pump.

Shipboard distilling plants are designed to produce distillate of very high quality. The chloride content of distillate produced must never exceed 0.065 equivalents per million. If the salinity of the distillate exceeds this value, an alarm sounds at the operating station, and a solenoid valve dumps the distillate to the bilges or overboard.

Even though properly operating distilling units are extremely efficient, these plants are not effective in removing volatile gases or liquids which have a lower boiling point than that of water. Accordingly, the OOD should inform the EOOW whenever the ship is operating in polluted waters, a situation which can exist even in mid-ocean if a ship immediately ahead pumps bilges or has an oil spill during refueling operations. Distilling units also are not effective in killing all microorganisms; thus the ship's senior medical representative conducts periodic tests of potable water samples to determine if chlorination is required for the ship's fresh water.

The following information concerning water production by the ship's distilling plant(s) is also of interest to the OOD:

In case of a pump failure, most distilling units can be "jury-rigged" to produce distillate of an acceptable quality, even though the "jury-rigging" may take some time to accomplish and will probably result in a reduced output of distillate. (Distilling units in *Spruance*-class DDs cannot be "jury-rigged," however.)

Steady operating conditions are essential to the satisfactory operation of any distilling unit. A fluctuation in the heat source, interruption of steam supply to the air ejector, or momentary loss of power to any of the pumps will affect the distilling unit. In extreme cases this could result in "salting up" the condensate phase of the propulsion cycle. Because steady operating conditions are essential to distilling units, engineer officers usually require that the output of the distilling plant be "sent to the bilges" during engineering casualty control drills.

While efficient operation of the ship's distilling plants is important to the OOD, he is most concerned with total water output. This brings us to the subject of "water management," a matter of concern to the Captain, the officer of the deck, the engineer officer, his EOOWs—and the entire ship's company.

WATER MANAGEMENT

Strictly speaking, the subject of water management has little to do with the electrical plant. But production on board the ship of each molecule of water depends to some degree on the electrical plant, and the intelligent use of water produced is of vital importance to the entire ship. Accordingly, water management is discussed here.

The engineer officer plans how water will be used on board ship with an overall policy that determines how much of the total output should be reserved for the (conventional) engineering plant, as well as how much water should be set aside as potable water for general use on board ship.

Water may be taken for granted when the distilling units are producing at full capacity and the Captain's Daily Fuel and Water Report shows a high percentage of feedwater and fresh water on board. Most OODs check those percentages before delivering the report to the Captain. Most of them realize that a conventional propulsion plant generally uses more feedwater at higher steaming rates than at slower speeeds. But many do not become overly concerned about the *amount* expended in each category, as long as the percentage on board remains high.

High feed usage in itself is alarming because it means that either serious leaks exist or else the feed or condensate systems are misaligned. High feed usage has a direct relationship to fuel usage because that water loss also represents a heat loss; relatively cold makeup feedwater must be heated so that the steam cycle can regain equilibrium. More fuel must be expended.

Hourly sounding of all tanks gives the EOOW an accurate account of how much water is on board. The EOOW uses water consumption figures to spot trends. Once he spots a trend, he begins an investigation of the cause of high water usage. Experienced officers of the deck know generally how much feedwater the plant uses at low, medium, and high speeds when the plant is "tight." They get a report of feedwater expenditure from the EOOW at periodic intervals.

Experienced OODs know that boiler blowdowns or "dumping" a boiler (replacing water in a boiler in an amount equal to its normal steaming level because of gross contamination of the boilerwater) naturally increases feed usage for a given period of time. Like the Captain, the engineer officer, and his EOOW, an experienced OOD is alert for any abnormally high feed usage. He knows the conventional propulsion plant must always have an adequate supply of feedwater.

If an excessive amount of potable water is being used and a distilling plant casualty or malfunction occurs, the overall water situation may quickly become critical. The OOD should ensure that water management policy is being followed, conferring with the Captain or engineer officer as necessary to determine if special conditions are to be placed into effect.

Most of "water management" concerns following guidelines, analyzing problems, and making decisions. The OOD will probably not be a "policy maker" concerning water management, but he may well be a key man in enforcing the policy put into effect. If necessary, men can forgo showers for a while. A full seabag for every member of the crew takes some pressure off the ship's laundry. Drinking water can be limited if sufficient juices and other liquids are available. But an engineering plant (and now, some cooling loops for many sophisticated weapons and electronic systems) always needs water.

During World War II, battle damage required some ships to use salty feedwater or even seawater for propulsion purposes so they could live to fight another day. Major boiler repairs were required after almost every incident. The higher temperatures and pressures in today's 1200 PSI propulsion plants make steaming with impure water an extremely dangerous practice.

Even marine gas turbines, the propulsion and generator prime movers for our jet age surface Navy—and the subject of the following chapter—require water for water washing to maintain efficiency.

Water in sufficient quantities is a necessity to all engineering plants. Its use must be managed.

CHAPTER REFERENCES

"Electric Propulsion Auxiliaries," Cdr. S. W. Krohn, USN, Bureau of Ships Journal, April 1966

Engineering, Operation and Maintenance, NavPers 10813

Principles of Naval Engineering, NavPers 10788-B

Shipboard Electrical Systems, NavEdTra 10864-D

8

Jet Age Propulsion

The U. S. Navy is the last of the world's major modern navies to "go gas turbine" as a means of propulsion for its surface warships. Significant numbers of heavily armed, gas turbine–powered warships such as the *Krivak*-class DDG exist in the Soviet naval inventory. In 1967 the Royal Navy made the decision to use gas turbine propulsion for all of its future surface warships. The French Navy, as well as some smaller navies, also has a number of well-armed destroyer and cruiser-type ships powered by marine gas turbines.

By the mid-1980s, nearly one of every three U. S. escort and ASW-type surface warships will be powered by gas turbines. *Spruance*-class destroyers and *Perry*-class guided missile frigates, which will account for the increase, both use General Electric LM 2500 marine gas turbines with controllable reversible pitch (CRP) propellers for propulsion. After 1985 the proportion of gas turbine–powered warships in the U. S. Navy is expected to increase even more, as older 1200 PSI warships such as the *Forest Sherman*–class destroyers come to the end of their active service life.

The engineers and officers on board the Navy's newest surface warships are justifiably proud of their ships and gas turbine propulsion plants. *Spruance*-class destroyers, for example, can have full power available from a cold iron status in less than 15 minutes. Destroyers of this class can accelerate from dead stop to speeds in excess of 30 knots in two minutes and from 10 to 30 knots in one minute. "Crash back," a maneuver that consists of moving from full power ahead to full power astern in a minimum amount of time, is an impressive maneuver on board any cruiser or destroyer but even more so on board *Spruance* because of the tremendous response of its propulsion gas turbines with their controllable reversible pitch propellers.

Some marine engineering experts contend that all FF/FFG/DD/DDG and nonnuclear cruisers of the future will be powered by gas turbines. Another group of designers points out that diesel engines, which also were developed after steam propulsion, did not replace steam for warship propulsion because of improvements in steam propulsion. Some of these men envision many surface warships of the 1990s using combination steam turbine and gas turbine propulsion plants. One of these plants is referred to as COSAG, which stands for Combined Steam and Gas (turbines). The Royal Navy already has warships with COSAG propulsion, in which the thermodynamic cycles of both prime movers are unrelated, although both prime movers can be mechanically connected to a common shaft. Another design is referred to as COGAS, which stands for Combined Gas and Steam (turbines). In this type of propulsion plant thermal energy contained in the exhaust of the gas turbine would be used to generate steam for a steam turbine; this thermodynamically related process would then result in both types of prime mover powering a common propulsion shaft.

However, let's leave any argument over future marine propulsion plant design up to the experts in order to consider now how this "jet age" propulsion is used in the U. S. Navy's first gas turbine–powered warships.

This chapter contains seven basic sections. The first concerns general information about gas turbine propulsion. Next is an abbreviated description of basic gas turbine operation. Following the section on basic operation is a more detailed look at gas turbine components and systems, which includes information on typical lubricating and fuel systems, the "bleed air" system and the relationship of bleed air to a typical start/ignition cycle.

Some elements of the propulsion power train are significantly different from the power train of surface warships driven by steam turbines described in previous chapters. Accordingly, the clutch/brake application to the power train is described, as is the controllable reversible pitch (CRP) propeller system and the relationship of integrated throttle control (ITC) to CRP control.

A following section requests the OOD or prospective OOD to mind some "P's" and "Q's" about gas turbine propulsion. Here we consider briefly such topics as marine gas turbine parameters, "perils," protection, and quality performance, which includes the turbine's responsiveness and dependability.

The last two sections are relatively brief. The first deals with a typical light off and underway sequence, which acquaints the reader with terms such as "dead shaft pick-up" and "plant mode logic." The final section lists propulsion plant casualties and possible effects on ship maneuverability.

Now, it's time to get under way with some general information about those gas turbines which will power many of our surface warships by the mid-1980s.

GAS TURBINE GENERAL INFORMATION

Here is some basic information about gas turbines. We first consider why the term "gas turbine" can be misleading to some Americans, and then look at details of gas turbine construction.

1. Because Americans often use the term "gas" in referring to gasoline, some students beginning their study of marine propulsion systems believe a "gas turbine" is some sort of prime mover that requires gasoline as its source of thermal energy. However, in a gas turbine the product of combustion is the working fluid, in this case a gas, used in the turbine blading in essentially the same manner that steam, which contains thermal energy, is the working fluid in a steam turbine.

2. A gas turbine consists of a *gas generator* and a *turbine which drives a load*. The gas generator consists of a compressor(s), a combustor (or combustion section), a gas generator turbine(s) (which drives the compressor), and an accessory drive section, which contains accessories required to control, operate, and lubricate the engine.

The turbine that drives the load is called the low pressure turbine, power turbine, or free turbine, depending on the manufacturer and the application. This turbine may be mechanically connected to the compressor as shown in the schematic diagram in Figure 8-1a; such a single-shaft gas turbine is often used as the prime mover in a ship service gas turbine generator (SSGTG), described in Chapter 7. However, in marine gas turbines used for propulsion, the power turbine and the free turbine, used in the LM 2500 and FT4 marine gas turbines, respectively, are only aerodynamically connected to the gas generator.

Schematic diagrams of the single-shaft; split-shaft (LM 2500) and twin-spool, split-shaft (FT4) gas turbines are shown in Figure 8-1.

3. Approximately 70% of all the power developed by the gas generator is required to drive the compressor (and gas generator accessories). Workable gas turbines would have been developed much sooner if anyone had known how to build a turbine that could develop enough power to drive the compressor and still have sufficient power left to deliver thrust, in the case of the turbojet, or torque, in the case of the marine gas turbine.

(An English engineer, Sir Frank Whittle, developed the first practical gas generator during World War II. Whittle's gas generator employed a centrifugal compressor, rather than the axial flow compressor used in today's turbojet, turboprop, turbofan engines, and the marine gas turbine. Design improvements in both compressors and the gas generator turbines that drive these compressors have resulted in the successful development

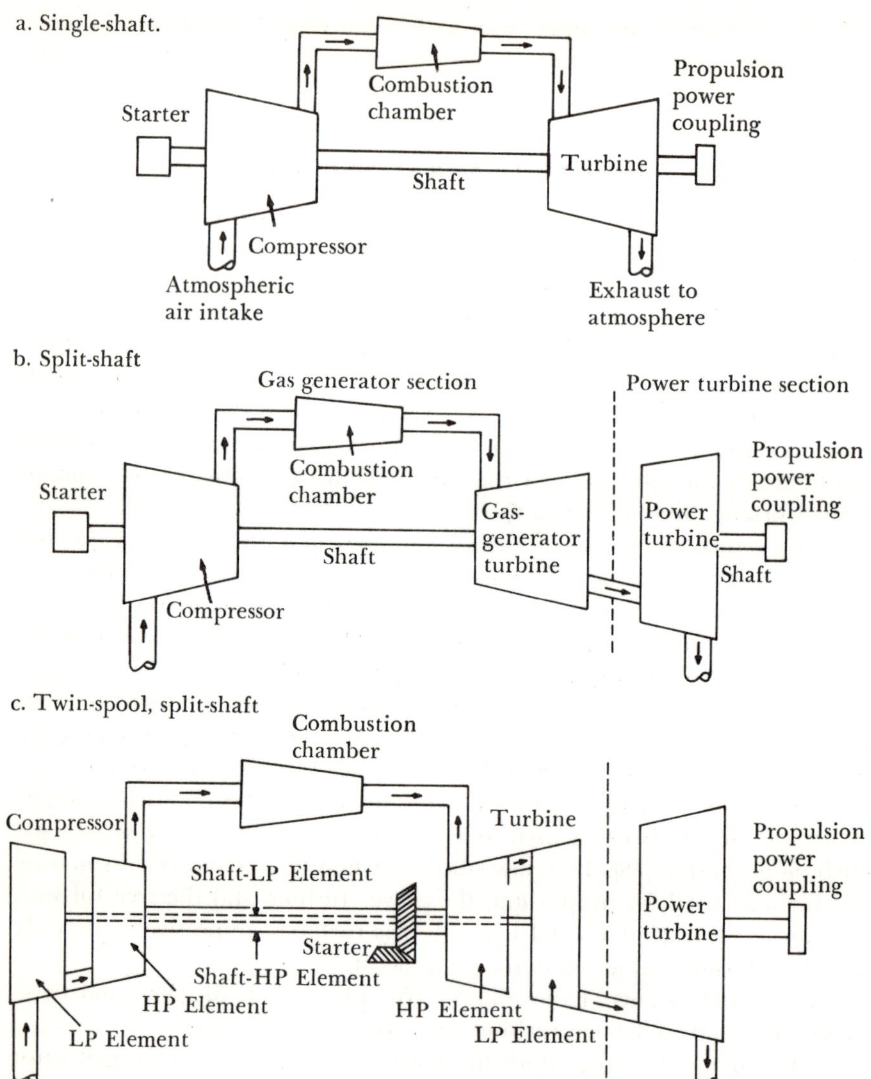

Figure 8–1. Schematic diagrams of basic gas turbines

of the turbojet and its family of engines, which includes the marine gas turbine.)

4. Steam used as the working fluid in conventional surface ship propulsion plants receives a tremendous amount of thermal energy during its manufacture in the boiler—but the exploding gas mixture that serves as the working fluid in a gas turbine is much hotter than even 1200 PSI steam at 950°F! The development of practical gas turbines would have

been impossible without the development of coatings and alloys that can withstand temperatures even higher than those in 1200 PSI propulsion plants. In addition, ingenious design features which allow air cooling of combustor elements, the gas generator turbine, and the power/free turbine also enable today's gas turbines to withstand high temperatures and stresses. Figure 8-2 is an illustration of how air is used to cool one stage of blading in the gas generator turbine of the LM 2500.

5. Our old friend, the temperature-pressure relationship, is as vital in gas turbine propulsion as it is in steam propulsion. Gas turbine engines operate on the *Brayton Cycle*. Figure 8-3 represents a Brayton Cycle for an unspecified but representative gas turbine. (Note the relatively small amount of energy available to the power/free turbine in a marine propul-

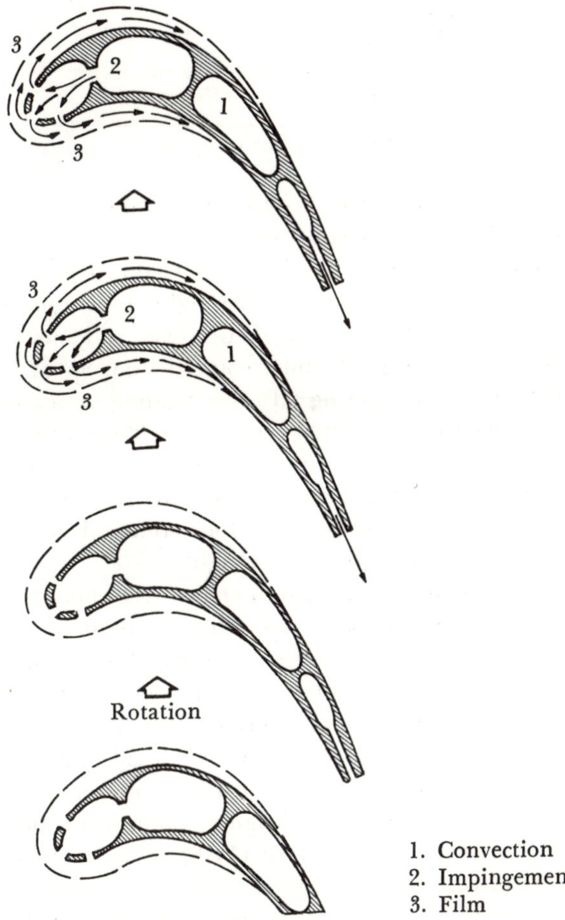

Figure 8–2. Air cooling of turbine blading

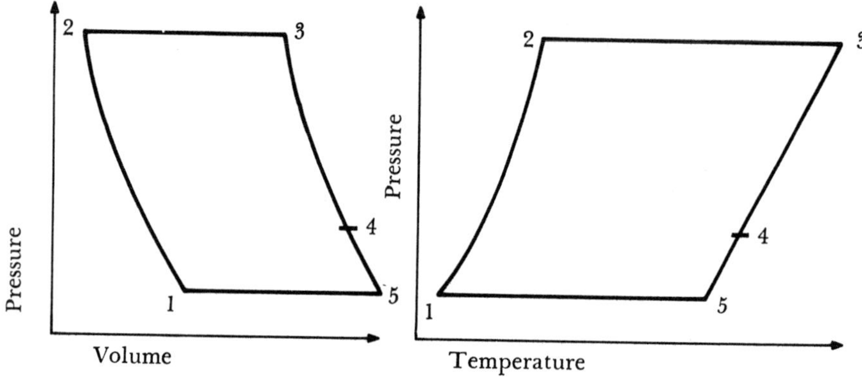

1. Atmospheric pressure/ambient temperature
1–2 Compression
2–3 Combustion
3–4 Expansion through gas generator turbine
4–5 Energy available to power/free turbine

Figure 8–3. The Brayton cycle

sion gas turbine, or to the thrust-producing section of a turbojet after expansion through the gas generator turbine.)

6. For the amount of power that it can develop, a gas turbine is an extremely lightweight prime mover. Lightweight, heat-resistant titanium alloys, which are among the metals used in rocket and satellite construction, are commonly used in gas turbine construction. Spool mounting of rotor blading, rather than mounting all rotor blading on a common shaft, saves weight. Hollow struts within the various turbine frame assemblies, which support the turbine and connect its sections, provide passage for lubricating oil flow, mechanical gearing, performance-monitoring sensors, and cooling air flow. The use of air flow for cooling throughout the gas turbine provides an additional savings in weight.

7. Obviously, gas turbines are precision engines. On board ship, propulsion gas turbines and SSGTGs are supported by numerous and rugged shock mountings to enhance battle survivability. But cleanliness of air, purity of fuel and lubricating oil, and cleanliness of the entire engine and ducting are also vital to gas turbines. Moreover, the precision and responsiveness of the gas turbine make the use of electrical and electronic controls a virtual necessity. On board gas turbine–powered warships the cause of more than one unexplained engine "shutdown" has later been traced to a faulty printed circuit board (PCB) or a less-than-clean electronic controlling device.

In the propulsion plant of a gas turbine–powered warship—perhaps more than in any other type of nonnuclear propulsion plant—a capital

"P" stands for Performance, Precision, and PMS (the Planned Maintenance System). Some Engineer Officers would add PQS, the Personnel Qualification Standard, which includes a basic understanding of how a gas turbine operates.

BASIC OPERATION OF A MARINE GAS TURBINE

The operation of a gas turbine is an expression of Sir Isaac Newton's Second Law of Motion, which states that "a change in motion is proportional to the force applied." This is expressed by the following equation:

$$F = Ma$$

In the gas turbine, the "M times a" is the mass of the gas generator exhaust gases multiplied by the amount this gas stream is accelerated while passing through the gas generator. In the marine gas turbine, the acceleration of the mass of gases through the gas generator results in a gas stream characterized by high velocities, temperatures, and pressures. The energy contained in this gas stream can be converted into mechanical energy by the power/free turbine.

Operation begins with rotation of the compressor by a starter. Air enters the compressor through a set of inlet guide vanes and is compressed at a ratio of approximately 1:1.2 in each stage of an axial flow compressor. Upon leaving the compressor, the compressed air is allowed to expand slightly in volume while passing through a diffuser in order to decrease its velocity and increase its static pressure before entering the combustor.

When a sufficient quantity of air has entered the combustor (combustion section of the gas generator), fuel is introduced in a very fine spray formed by the manifold/fuel nozzle configuration of the combustor. Ignition occurs when spark igniters similar to sparkplugs in an automobile are activated. The igniters are turned off when the gas generator is operating, since combustion will be sustained as long as there is a suitable mixture of air and fuel in the combustor.

The addition of thermal energy through combustion rapidly accelerates the mass of air, and this working fluid consisting of the gases of combustion rushes toward the gas generator turbine. (The Second Law of Thermodynamics, referred to in Chapter 1, is also now "in on the act.") The gas generator turbine, referred to as the "high pressure turbine" in the LM 2500 gas turbine, converts energy contained in the gas stream to mechanical energy to drive the compressor. Approximately 70% of the energy contained in the gas stream after combustion is required to drive the compressor and gas generator accessories. The energy remaining after expansion of the gas through the gas generator turbine will be available to the power/free turbine, which can be clutched to the propeller shafting for ship propulsion.

After expansion through the gas generator turbine, the gas stream passes through another diffuser, which slightly reduces velocity and allows static pressure to increase. Then the gas stream is expanded through the power/free turbine. Despite the fact that a majority of the energy contained in the gas stream was used in the gas generator turbine, a tremendous amount of energy is still available for conversion to additional mechanical energy by the power/free turbine. As an example, the temperature of the gas stream at the power turbine inlet of the LM 2500 can be as high as 1500°F; the energy contained in the gas stream will enable the LM 2500 to develop a maximum brake horsepower rating of 21,500 BHP for propulsion in a *Spruance*-class destroyer (20,500 BHP for a *Perry*-class FFG).

Fuel is metered according to a predetermined schedule which permits the gas generator to produce exhaust gases (the gas stream) that contain the energy required by the power/free turbine to respond to all load demands. After expansion through the power/free turbine, the gas stream is exhausted to the atmosphere via the gas turbine exhaust ducting. A considerable amount of thermal energy is contained in the gas stream even after expansion through the power/free turbine; the amount is large enough to present a distinctive infrared (IR) signature to an enemy force that possesses IR detection sensors.

GAS TURBINE COMPONENTS AND SYSTEMS

Sections of Chapters 1 through 4 contain brief descriptions of equipments and systems required to support conventional surface ship propulsion. The officer of the deck of a gas turbine–powered warship shares common responsibilities and concerns with his counterpart in a steam turbine–powered warship. Knowing *how* his basic propulsion plant operates is vital to any OOD or prospective OOD. Accordingly, this section will briefly describe gas turbine components and systems required for gas turbine propulsion, such as the fuel oil service system, the lubrication system, the bleed air system (BAS), and a marine gas turbine start/ignition cycle.

The goal of this and subsequent sections is to enable the OOD to understand how his propulsion plant partner, the EOOW, and the remaining members of the mobility team control their marine gas turbine propulsion plant.

The Compressor

The secret of why a gas turbine operates lies in the compressor. Depending on the design of the particular gas turbine, the gas generator turbine extracts 65–75% of the total energy in the gas stream after combustion in order to drive the compressor. But the work the compressor does in compressing air for combustion is necessary, since a gas stream resulting from the combustion of fuel and air at atmospheric pressure would not expand enough to do useful work. The energy required for compression will be regained after combustion, as the rapidly accelerated air mass (gas stream) roars out of the combustor with a tremendous level of energy contained in it. The amount of power a marine gas turbine can develop is directly related to the efficiency of its compressor.

The great majority of marine gas turbines use axial flow compressors rather than centrifugal flow compressors. Centrifugal compressors are easier to design and manufacture and are generally more durable than axial flow compressors, but axial flow compressors require much less frontal area for a given compression ratio. The LM 2500 compressor is a 16-stage, axial flow compressor which compresses the air in a ratio of approximately 16:1. Each stage of an axial flow compressor consists of a *stator vane* and a *rotor blade*. Unlike the turbines in a gas turbine, which also consist of stages composed of stator vanes and rotor blades, the flow path through an axial flow compressor decreases in cross-sectional area, in proportion to the reduced volume as the air is compressed in progressive stages.

While the compressor's primary function is to compress and pump air for combustion, not all the air will enter the combustor. Some of the air will be drawn off at various stages, depending upon compressor design, for engine cooling, seal pressurization of the various sumps in the "dry sump" lubrication systems employed in gas turbines, and "bleed air" for ship service use. Actually, not all the air that enters the combustor is utilized for combustion. Approximately 75% of the air that enters the combustion chamber will be used to "center" the flame pattern and to cool the liners of the combustion chamber.

Although gas turbines are much more efficient now than those developed immediately after World War II, improvements have brought problems. *Compressor stall* is a problem the engineers in a gas turbine warship are concerned with, which the leading engineman or gas turbine specialist could probably explain to you somewhat like this: "Think of the blades in our axial flow compressor as airfoils—which they are—rather similar in shape to the cross section of an aircraft wing. If the aircraft pilot performs a maneuver that disturbs the smooth airflow over his wing, the wing may lose a substantial portion of its lift. An unstable condition occurs which could have serious consequences, and the pilot has to react quickly to restore the *angle of attack* to one which will permit that smooth airflow to resume."

Figure 8-4 is a schematic diagram illustrating angle of attack and stall.

Stalls may occur in the compressor of a marine gas turbine because of a change in air density at the bellmouth inlet, distortion of the air flow pattern through the compressor, or even a small blockage of the air supply to the compressor. Stall is most apt to occur during rapid acceleration.

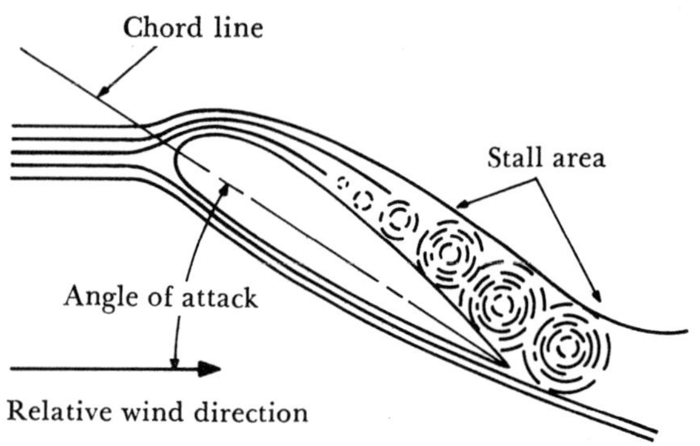

Figure 8–4. Schematic diagram of angle of attack and stall

Stalls vary in intensity from a low engine rumble, in the case of a partial stall, to a violent explosion in the case of a complete stall.

Some compressor stalls are only transient and minor in nature. Other compressor stalls could have damaging effects. In *Spruance*- and *Perry*-class warships, propulsion plant operators who note a rapidly rising or extremely high temperature at the power turbine inlet, or failure of the gas generator to accelerate, recognize that a compressor stall with potentially serious consequences may result. Under such conditions—which the operator would term "high T5.4" or "low N_1," to denote high power turbine inlet temperature or failure of the gas generator to accelerate, respectively —the operator would initiate engine "shutdown" procedures. If either condition exceeded normal operating parameters with gas generator speed in excess of 7500 RPM, the engineer officer would report the propulsion gas turbine as being out of commission to the Captain. Such a casualty would require "borescoping" various components of the gas turbine to see if major damage had occurred.

Fortunately, design features of modern gas turbines virtually eliminate compressor stalls. In the LM 2500 the inlet guide vanes (IGV) and first six stages of compressor stator vanes can be positioned to ensure a smooth air flow through the compressor, even under conditions of rapid acceleration and high compressor inlet temperature (which equates to low air density). These variable stator vanes, or "VSVs," as some of the engineers in *Spruance*- and *Perry*-class warships like to call them, are positioned by hydraulic actuators operated by fuel pressure from the main fuel control (MFC). (The MFC is a fairly complex device that receives inputs such as compressor inlet temperature, compressor discharge pressure, gas generator speed, fuel specific gravity, and desired throttle setting to prevent overtemperature and stall during acceleration and loss of combustion, or "flameout," during deceleration.)

In marine gas turbines such as the FT4, which is used for propulsion in the Coast Guard's *Hamilton*-class cutters, stall is defeated by the use of dual or twin-spool compressors. In marine gas turbines such as this, RPM of the rear, or high pressure, compressor rotor is governed by fuel control. The front, or low pressure, compressor rotor is rotated by its turbine at a speed that will ensure optimum flow through the low pressure compressor. Such a configuration of low and high pressure compressors ensures optimum compressor operation throughout the operating range of the gas generator. Other marine gas turbine axial flow compressor designs feature the use of valves to bleed air from initial compressor stages in conditions under which stalls could be encountered. This prevents excessive pressures from developing in the latter stages, which would overload the compressor and result in stall.

The Combustor

Gas turbine designers say the combustor, also referred to as the "combustion chamber," is the most efficient component of a gas turbine. Depending upon design, the combustor has an efficiency of 95–98% over wide operating ranges.

The following are requirements for all three basic types of combustion chambers: uniform flame pattern, good flame stability, light weight, reliability, resistance to the formation of carbon, and a reasonable length of life.

The can-type combustion chamber, used with both axial and centrifugal flow compressors, is the simplest of all the combustors, consisting of a highly heat-resistant, perforated liner within an outer housing. Each can-type chamber has an individual air inlet duct.

An annular combustion chamber, such as that used on the LM 2500, has a cross-sectional area similar to that of the latter stages of the compressor and the initial stage of the gas generator turbine. Fuel is introduced through a series of nozzles connected to a common manifold and mounted equidistant from each other at the front of the combustion chamber. Diffusion of air and an efficient flame pattern result from rows of holes punched in the burner outer liner (also called the "basket").

Can-annular combustion chambers consist of cans mounted axially to allow an annular discharge of air from the compressor. These combustion chambers allow removal of individual cans for ease in maintenance. Individual cans within the removable steel shroud that covers them are interconnected by means of projecting flame tubes.

As mentioned previously, only about 25% of the air that enters the combustor is used for combustion. However, combustion of this *primary air* and fuel produces a gas that can have a temperature as high as 3500°F. This gas must be cooled to approximately half that value before encountering the turbine(s). Cooling this gas and protecting the combustor from overtemperatures are the function of *secondary air,* which is admitted at downstream points in the combustion chamber. The resulting gas stream, which contains tremendous amounts of thermal energy, is the working fluid that roars toward the gas generator turbine.

The Gas Generator Turbine

The major purpose of the gas generator turbine is to drive the compressor fast enough to produce the amount of air required to keep combustion self-sustaining. The task sounds simple, but the requirements placed upon the gas generator turbine are demanding: It must extract roughly 70% of the energy from the gas stream to perform this function. This will subject the gas generator turbine to temperatures that can exceed 1800°F and high rotational speeds which impose severe centrifugal loads on the rotor and its blading.

Development of gas turbines as prime movers was delayed until suitable metals could be developed to withstand the extreme temperatures and forces placed on the turbine. Designers made further contributions through the ingenious design of modern gas turbine blading; each blade performs as an impulse blade at its root and a reaction blade at its tip, thereby balancing the load on the blade throughout its length. (The blades of a modern gas turbine are significantly longer than and different in shape from either the impulse blades or reaction blades in similar stages of a steam turbine as described in Chapter 4.) Practical methods of using air flow to cool turbine blading and entire components of the prime mover were necessary in the development of gas turbine propulsion. Still, the condition of turbine blading is generally accepted as the governing factor determining the length of time required between normally scheduled overhaul/engine changeout.

Each stage of a gas generator turbine is composed of a stator vane and a rotor blade. The turbine converts thermal energy contained in the gas stream to mechanical energy in much the same manner that the steam turbine described in Chapter 4 does. Stages are progressively larger to permit efficient extraction of energy at reduced velocities in downstream stages.

Shrouding at the tips of blades in some turbine installations is used to prevent excessive vibration and distortion of the blade under high loads. (Distortion or elongation of the blade or vane is known as "creep." Creep is cumulative, with its rate being determined by the load on the turbine and the strength of the blading, which is in turn influenced by the temperature within the turbine.) Shrouding allows the tips of the blades to touch each other and has an aerodynamic advantage in that thinner blades can be used. However, shrouding requires that turbines run at lower RPMs and that adequate cooling be provided because of the extra mass at the blade tip.

Gas generator turbines are of either the single-shaft or the twin-spool variety. Single-shaft gas turbines can also be connected to the load, as in the case of the Allison 501 gas turbine. (This gas turbine is the SSGTG on board *Spruance*-class warships.)

Regardless of its type or the manufacturer, the gas generator turbine is one of the gas turbine components most susceptible to damage because of the heat and forces it must withstand. Propulsion plant operators monitor operation of the gas generator turbine closely to ensure that normal parameters are not exceeded.

The Accessory Drive Section

The final component of the gas generator is the accessory drive section, which performs two primary functions: (1) it provides space for the mounting of accessories necessary for the operation and control of the gas turbine, and (2) it acts as an oil sump/reservoir and provides housing for the reduction gearing and drive for the control accessories.

The accessory drive section, including the transfer gearbox, is normally located under the compressor of the gas generator. Rotational power is provided by the gas generator turbine rotor through the various reduction gears.

Basic accessories in the accessory drive section include the fuel control with its governing device, a fuel oil pump, the lubricating oil sump, the lube oil and scavenging pump, an auxiliary fuel pump, and a starter.

The Power/Free Turbine

The power, or free, or low pressure turbine—depending on what the manufacturer calls it—is the final component of the marine gas turbine used for propulsion.

The power/free turbine is *not* mechanically connected to the gas generator turbine rotor. Instead, it is aerodynamically connected to the gas generator by the output of the gas stream remaining after expansion through the gas generator turbine. Aerodynamic coupling enables the power/free turbine to respond efficiently and rapidly to changing load conditions. For propulsion purposes, the power/free turbine is mechanically coupled to the propeller shafting—but more about that later.

The physical dimensions of stages in the power/free turbine are much larger than those in the gas generator turbine, since the gas generator turbine has already converted most of the energy contained in the gas stream to mechanical energy. Power/free turbine stages must be correspondingly larger to handle the higher volume of gas flow, which now contains only about 30% of the energy of the original gas stream. In most modern gas turbines the physical appearance of blading in both turbines is similar, except that the power/free turbine blading is larger. The various stages of power/free turbines also consist of stator vanes and rotor blades. As in the case of the gas generator turbine, cooling for the power/free turbine is provided by air bled from the compressor.

A high-speed coupling shaft, splined and bolted to the power/free turbine, extends through the exhaust duct. This shaft provides the means for eventual connection of the prime mover with the load.

Gas Turbine Bearings and Lubrication

Gas turbines are precision prime movers that develop—and are exposed to—extremely high temperatures and forces. The bearings that support gas turbine components and the lubricating system which cools these bearings are critical to the performance and life of the engine.

Ball and roller bearings are commonly used in marine gas turbines. Figure 8-5, a schematic drawing displaying the location of bearings in the LM 2500, is representative for marine gas turbines in general, although the number of bearings and numbering system may differ among manufacturers.

In Figure 8-5, the bearing numbers 1 and 2 are not used. Bearing numbers 3, 4R, 4B, and 5 are used in the gas generator. Bearing numbers 6, 7B, and 7R are associated with the power turbine, with 7B mounted on the rear shaft of the power turbine and responsible for carrying the thrust load of the power turbine rotor.

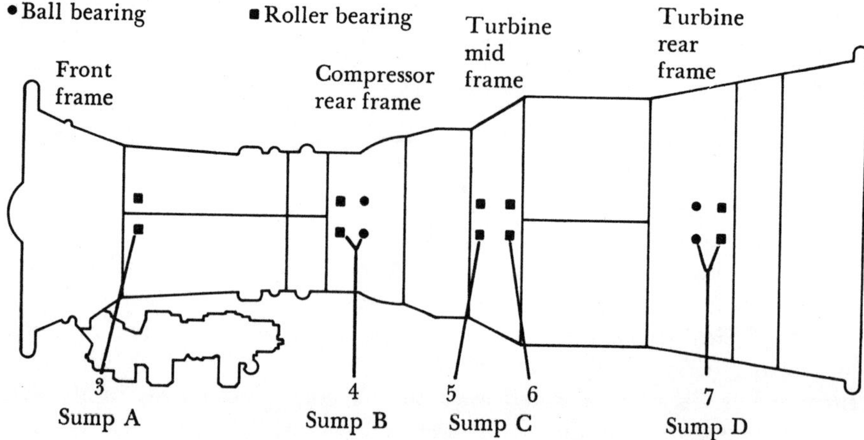

Figure 8–5. Bearing location and numbering in LM 2500

Normal lubricating oil such as the 2190 TEP used to lubricate bearings in the high and low pressure turbines and the main reduction gears of a conventionally powered surface warship would quickly lose its lubricating qualities in a gas turbine. The bearings, gears, and splines in a marine gas turbine require *synthetic* lubricating oil to withstand the extremely high temperatures associated with the fuel combustion and high rotation rate of gas turbine elements. (However, the main reduction gears of *Spruance*- and *Perry*-class warships are basically similar to the main reduction gear installation of a conventionally powered surface warship, and 2190 TEP from the main reduction gear lubricating system in *Spruance*- and *Perry*-class warships serves as the cooling medium for the synthetic lube oil used in the propulsion gas turbines of these warships.)

A summary of lubrication of the propulsion gas turbines of *Spruance*- and *Perry*-class warships is as follows:

Synthetic lube oil flows from the lube oil storage and conditioning assembly—also known as the "LOSCA," another of those exotic terms used by the engineers on board these ships—to the supply element of the lube and scavenge pump. Then the oil passes through a duplex filter and a check valve, and into the supply manifold. From the manifold the oil flows to the four bearings and the transfer gearbox of the accessory drive section. The oil is then drawn from the bearing sumps and the transfer gearbox by the five scavenge elements of the lube and scavenge pump for return to the LOSCA.

Lube oil that escaped from the bearing sumps would be a definite fire hazard inside the gas turbine, which could result in a catastrophic turbine casualty. To prevent this, air drawn from the eighth stage of the LM 2500 compressor is used to pressurize a cavity between oil seals in each lube oil sump. Use of this feature and a scavenge element to suck oil supplied to the bearing in the sump creates a "dry sump" system, which is the predominant method of lubricating bearings in today's marine gas turbines.

Basic Propulsion Gas Turbine Fuel System

Although fairly complex, fuel systems for propulsion gas turbines are highly reliable systems designed to regulate and distribute fuel to the combustor in the proper amount to control gas generator speed.

In the basic fuel system, the engine-driven pump in the accessory drive section receives fuel from one of the two motor-driven fuel oil service pumps in the machinery room via the fuel oil supply line and a duplex strainer. The engine-driven pump boosts the fuel pressure and forces the fuel through a high-pressure filter to a fuel control governor device in the fuel control assembly. The fuel governor device provides fuel to the combustor nozzles at the desired pressure and volume to control gas generator speed. A nozzle shut-off valve downstream of the fuel control assembly provides a positive means of stopping fuel flow to the combustor if a dangerous condition develops.

Since there are a number of different types of gas turbines used for propulsion, the fuel systems for ship classes vary. OODs and prospective OODs of *Spruance*- and *Perry*-class warships may find the following points helpful in achieving a basic understanding of the fuel system for the LM 2500 propulsion gas turbines in their ships:

- The fuel pump contains both a *boost element* and a *high pressure* element, as well as a filter, a filter bypass relief valve, and a high pressure relief valve.
- The *main fuel control* (MFC) is located immediately downstream of the fuel pump. The MFC controls gas generator speed and schedules acceleration, deceleration, and minimum fuel flow during start-up. The MFC also schedules the position of the variable stator vanes, based on gas generator speed and compressor inlet temperature. (Review of the section on compressor stall in this chapter may be helpful in achieving a basic understanding of the role of the MFC.)
- A *pressurizing valve* immediately downstream of the MFC ensures that the fuel system has adequate fuel control servo supply pressure and variable stator vane actuation pressure, even during low gas generator speed demands.
- An electrically operated fuel *purge valve* is employed in the system to drain low temperature fuel prior to gas turbine start-up (a safety feature). This valve can be operated from either the central control station or the local operating station when a purge is required or by control logic during automatic sequence starting.
- *Two fuel shutdown valves* just before the fuel manifold-nozzle configuration prevent fuel from entering the combustor, thereby providing the same quick-closing, safety shut-off feature that the quick closing valve (QCV) described in Chapter 3 provides. These valves are operated in parallel by control logic during automatic sequencing at the local operating station or manually from either the central control or the local operating station in manual or manual initiate sequencing. The second valve acts as a back-up if the first fuel shutdown valve fails to function during a shutdown condition.

Marine gas turbines are efficient prime movers that can operate on a number of distillate fuels of various grades. However, the fuel system must be adjusted for the *fuel specific gravity* of the particular batch of fuel available to the gas turbine.

The Bleed Air System

The BAS is a compressed air system that uses air extracted from the compressors of marine propulsion gas turbines (and the SSGTGs in

Spruance-class DDs) to perform certain vital tasks in gas turbine–powered warships. The BAS serves as the following:

1. A source of starting air on gas turbines that are air-started (this includes LM 2500 gas turbines).
2. The source of anti-icing air for marine gas turbines.
3. The source of both Prairie and Masker Air on board *Spruance*- and *Perry*-class warships. (The Prairie Masker Air System on board *Spruance*- and *Perry*-class warships is basically similar to the Prairie Masker System on board 1200 PSI FFs described in Chapter 5, with the exception that Prairie Masker air in these ships is supplied by steam-driven air compressors.)

The BAS consists of bleed air outlet valves on each turbine, pressure-regulating and control valves, the bleed air header, distribution piping, and bleed air control and monitoring circuitry. (Coolers for both Prairie and Masker air which utilize seawater as the cooling medium are also part of the BAS in *Spruance*- and *Perry*-class warships.) Even though the pressure of bleed air in the headers and distribution piping is only about 75 psi, the temperature of the bleed air is in excess of 500°F. As a result, these lines must be lagged for safety of personnel and to decrease fire hazards.

If ice-forming conditions exist—a situation that could lead to gas turbine damage and will be explained later—bleed air must be used to heat the air entering the compressor. Alarms will indicate that ice-forming conditions exist, and anti-icing air can be enabled by the propulsion plant operators. A requirement to combat ice-forming conditions results in a considerable expenditure of bleed air.

A considerable expenditure of bleed air also results when starting more than one gas turbine simultaneously.

Marine Gas Turbine Start/Ignition Cycle

Most marine gas turbines, including the LM 2500, are started by rotating (motoring) the compressor with compressed air supplied from an external source. In the case of the LM 2500, the normal source of starting air is bleed air mixed with cooler Masker air, which is directed to the starting piping by the Masker air transfer valve and then introduced into the turbine to begin compressor rotation. (If insufficient bleed air is available for starting, HP air from nearby flasks is reduced in pressure and made available for starting.)

When the compressor speed reaches a designated RPM, which is sufficient to purge any unburnt fuel or explosive fuel-air mixture from the gas turbine, ignition takes place. With combustion, compressor speed increases, and a centrifugal switch deactivates the starter. The ignition system is aslo deactivated at this time.

The following types of abnormal start may occur:

1. *False start:* no start occurs, as a result of either fuel system or ignition system problems. Before a restart is attempted, the gas generator must be motored to ensure that all explosive mixtures are purged.

2. *Hot start:* a loud, rumbling noise, flames from the exhaust stack, and high temperature indications! The engineer officer will request permission from the Captain to place this gas turbine out of commission and borescope it to see if damage to any component has resulted.

3. *Hung start:* ignition occurs, but the gas generator takes an excessively long time to achieve idle speed. The engineers will consult their technical manuals about what to do in the case of this abnormal start, and the EOOW will take little time in making a recommendation to the bridge.

In this discussion, however, let's assume that normal start has been achieved. Now it's time for a look at the propulsion plant's power train, including the method of clutching the prime mover to the power train and a brief description of the CRP propeller system. You'll learn that ITCs—which stands for integrated throttle controls—"work" well with CRPs. And don't go away—even more exotic terms, such as "FOD" and "dead shaft pick-up," will be encountered in later sections of this chapter.

PROPULSION POWER TRAIN

Reduction Gears

As in the case of steam turbine propulsion, reduction gears are used in gas turbine–powered warships to match the speed of the prime mover to the speed of the propeller.

Reduction gears such as those on board *Spruance*- and *Perry*-class warships are quite similar to main reduction gears referred to in Chapter 4. The main reduction gear lubricating system, consisting of two lube oil service pumps and an attached pump that supplies sufficient pressure by itself at higher speeds, is also similar to that of conventionally powered DD/DDG/FF/FFG and CGs. Significantly, loss of reduction gear lube oil pressure or a casualty characterized by a loud, roaring noise in the reduction gears poses the same threat to the mobility of a gas turbine–powered combatant as either casualty would to a conventionally powered surface ship. Such a casualty calls for the affected shaft to be stopped immediately as a normal procedure. However, the OOD must know the Captain's Stop Shaft Policy in situations involving restricted maneuvering, since it may differ from normal stop and lock procedures.

The major difference in the power train connection of steam turbine warships and gas turbine warships in general is in the method of connecting the prime mover to the gear train. Clutches are used in gas turbine–

powered warships. (Ships powered by diesel engines also use clutches for power train connection.)

Reversing clutches enable a ship with a prime mover that rotates in only one direction to move either ahead or astern. CRP propeller systems —in which the angle of the propeller blades can be varied to produce changes in speed as well as a change in direction—also enable a ship to move either ahead or astern without declutching the prime mover from the gear train.

Spruance- and *Perry*-class warships use CRP systems with an air-actuated *clutch/brake* assembly for each engine to connect their propulsion gas turbines to the propulsion power train. (*Newport*-class LSTs, which are powered by diesel engines, also use a similar power train connection with a CRP system for propulsion.) Since the majority of the Navy's gas turbine–powered warships for years to come will be *Spruance*-class DDs and *Perry*-class FFGs, brief descriptions of their clutch/brake gear train connection and their CRP system will be given in this section.

Clutch/Brake Assembly

Figure 8-6 is a schematic diagram showing major components in the port propulsion system of a *Spruance*-class DD. (Both shafts are inboard-turning shafts in this class ship, which gives the ship good steering control when going astern—but not much twisting power in restricted waters.)

The clutch/brake assemblies are intricate but reliable assemblies in which the clutching arrangement and the brake function independently of each other.

The clutching arrangement is air-operated, cooled by reduction gear lube oil, and consists of two separate clutches. To connect the high-speed coupling connected to the power turbine of a propulsion gas turbine, with the first reduction pinion (high speed pinion) of the main reduction gear, propulsion plant operators would push the "clutch engage" button at either the central control or the local operating console. If all "permissives" were met, air would be ported through the assembly and force a *friction clutch* against the first reduction pinion shaft. When the input shaft from the power turbine and the first reduction pinion were within 11 RPM of each other (another permissive), air pressure would force a dental clutch pushrod to move forward, enabling the *dental clutch* to engage a sleeve gear in the first reduction pinion. Engagement of the dental clutch with the high speed pinion permits the power turbine to transmit the high torque it is capable of producing to the propeller shafting via the high speed coupling and the reduction gears.

(In your conversations with the engineers in gas turbine–powered warships, you may hear the word permissive quite often. Gas turbine propulsion has its own terminology, as do the control systems that are used

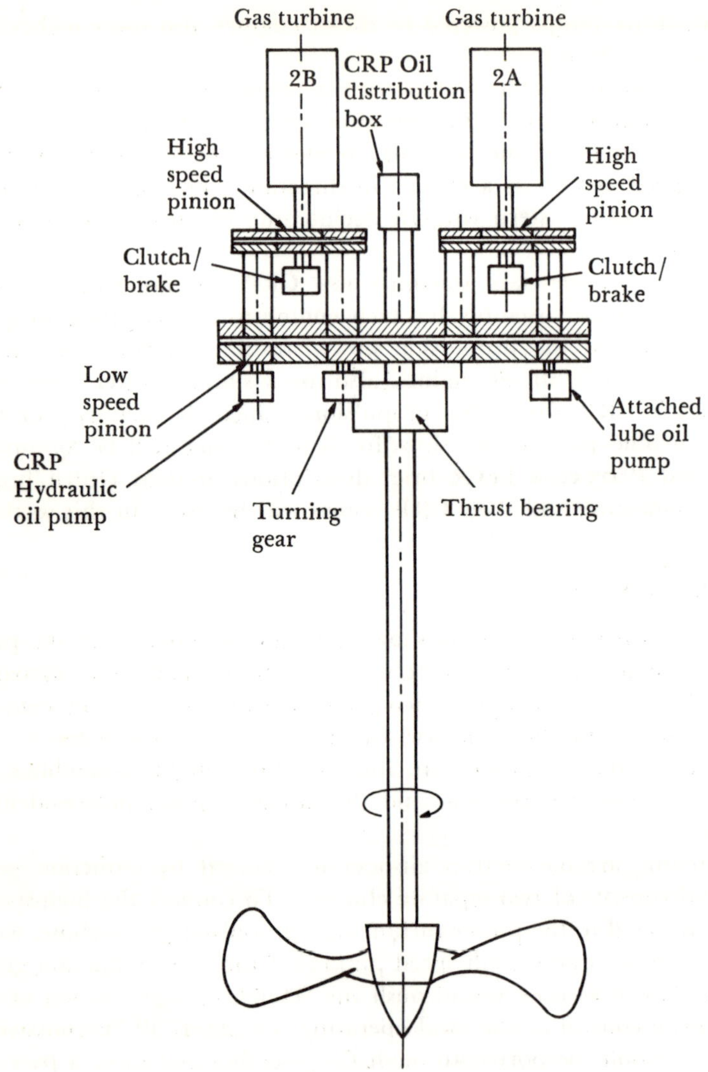

Figure 8–6. Major propulsion power train components of a *Spruance*-class DD (port shaft)

with it. Just about the time you realize that PAMCE stands for propulsion [and] auxiliary machinery control equipment, which is located in CCS—the central control station, where the EOOW exercises control of the propulsion plant—you'll hear the acronym PLOE, for propulsion local operating equipment, which includes the console where an operator can take control of the gas turbines located in that engineroom. Then the term permissive may come up. Basically, a permissive is a condition that has to exist before the control logic used to maintain control of marine gas turbines will permit a subsequent event to take place. That subsequent event may itself be a permissive. It is important to remember that even though a number of permissives may have to be satisfied before the desired result is achieved, controls associated with gas turbine propulsion permit these final results to be achieved rather rapidly.)

The clutch disengagement sequence is the reverse of the engagement sequence.

The brake assembly of the clutch/brake can be used either for stopping the power turbine, in which case the clutch must be *disengaged,* or for propulsion shaft braking, in which case the clutch is *engaged.* (However, the brake is not to be used to stop the propulsion shaft in the event of a loss of reduction gear lube oil pressure. The reason for this is that the clutch/brake is lubricated and cooled by reduction gear lube oil, and possible "wiping" of the clutch/brake could result in contamination of the reduction gear lube oil system.) The brake is also actuated by air, and consists of rotating and stationary discs which are forced together to prevent rotation of the power turbine input shaft when air is admitted to the brake from the COAS (control air system). A separate COAS is provided for each clutch/brake assembly and mounted on either side of the reduction gear casing.

A loss of reduction gear lube oil pressure and a casualty characterized by a loud, roaring noise in the reduction gears pose tremendous threats to the propulsion power train of any warship. Here's how the EOOW in a *Spruance*-class destroyer would attempt to combat these casualties (assuming the ship is not in a restricted maneuvering situation covered by the Stop Shaft Policy of the commanding officer):

In the case of loss of reduction gear lube oil pressure, the EOOW would take manual control of both throttle and pitch and attempt to achieve a *zero*-thrust, PLA-*idle* condition as quickly as possible. When this condition was achieved and other permissives were satisfied, the EOOW would then apply the brake to hold the propulsion shaft for locking. (How the reduction gears are locked by engaging the locking feature of the turning [jacking] gear is explained in Chapter 4.)

If the ship was making high speed when noise in the reduction gears occurred, the EOOW would attempt to lower shaft RPM and decrease pitch to a point at which the shaft brake could be applied with a degree of certainty that the propulsion power train could be stopped. He would do this as quickly as possible. The reduction gears would be locked as quickly as possible in this type of casualty, just as in the case of loss of reduction gear lube oil pressure.

In either case—a loud, roaring noise in the reduction gear or loss of reduction gear lube oil pressure—the EOOW would also request the OOD to lower speed on the unaffected shaft if the tactical situation permitted. This would reduce "drag" on the affected shaft, which must be reduced as much as possible during the precision shaft locking evolution.

Loss of air pressure supplied by the COAS has the following effects on the clutch/brake assembly: (1) the brake will *disengage,* and (2) the friction and dental clutches will be either *engaged* or *disengaged,* depending on the position the clutches were in when the air loss occurred.

Control of the clutch/brake assemblies allows different propulsion plant configurations to be selected, as follows:

Secure	Both propulsion gas turbines associated with the power train are disengaged.
Split plant	Either the A or the B propulsion turbine is engaged to the power train.
Full power	Both propulsion gas turbines are engaged to their respective power train.

The clutch/brake assemblies allow plant mode changes to be made without interrupting power to the propulsion shaft. This includes switching propulsion gas turbines to remain in the split plant mode.

Controllable Reversible Pitch (CRP) Propeller System

Controllable, reversible pitch propeller systems, which are not new to the Navy, provide ships with excellent maneuverability. A ship with CRP propellers requires much less distance to stop than a ship with fixed-pitch propellers.

CRP propellers are used extensively in minesweepers. CRP propellers enable landing ships such as ships of the *Newport*-class LST to maintain a definite position in relation to the shoreline, and further permit these ships to take advantage of the high degree of maneuverability they must have to retrieve beaching gear and clear the landing area.

In the case of *Spruance*-class destroyers, a combination of excellent rudder response, low self-noise signature as a result of the Prairie Masker System, and rapid acceleration/deceleration capability resulting from gas turbine propulsion and the CRP system gives these ships the ability to conduct a "close-in" ASW action with a high degree of proficiency.

A brief description of the CRP system used in both *Spruance*- and *Perry*-class warships follows:

Forward and astern thrust is accomplished by changing the pitch of the CRP propeller from 100% pitch ahead (about 26′) to 49% pitch astern (about 13′). The mechanism that positions the angle of the propeller blades is located in the propeller hub. The hydraulic oil (2190 TEP, the same type used in the reduction gear) to operate this mechanism is supplied by the HOPM (hydraulic oil power module), which is located near the reduction gear. The HOPM delivers oil to the OD (oil distribution) box, located at the forward end of the reduction gear's bull (main) gear shaft, to be used in two different ways: (1) *high pressure oil* is directed from the OD box to the propeller hub through tubing contained within the propeller shaft, and (2) *control oil* is used within the OD box to actuate propeller pitch changes. (This results in the pitch change in the OD box being transmitted to the positioning mechanism in the propeller hub by the high pressure oil.)

Pitch is normally controlled with the ITC on the bridge or in the central control station. Pitch can also be controlled from the local control console in the engineroom. A casualty in the normal mode of operation could require that pitch be controlled manually at the OD box.

For emergency operation, pitch can be set at 100% pitch ahead and locked independently of the hydraulic system. The engineers might call this the "emergency/take home" feature of their pitch system, and they would use shaft RPM to vary speeds from about 12 to 20 knots on that shaft. (Higher speed would be dangerous because the oil in the OD box is no longer allowed to circulate for cooling purposes.) Pitch could also be set astern if necessary independently of the hydraulic system (but not locked).

The OOD or prospective OOD of a *Spruance*- or *Perry*-class warship should remember four other things about his CRP system:

1. Loss of oil pressure in the CRP system will normally result in the blades assuming an astern pitch configuration.
2. The maximum rate of pitch change from full ahead to full astern requires about 30–35 seconds.
3. A 0% pitch indication rarely results in a "no-thrust" condition. In many ships a 4–8% pitch ahead indication is required to produce *zero thrust*. Even this value may change over a period of time, which is one reason why the Captain and the OOD are peering over the side of the ship and checking mooring lines so anxiously while conducting tests of the propulsion system prior to getting under way.
4. Before entering port or beginning any evolution involving restricted maneuvering, the pitch control system response should be checked by decreasing pitch slightly and noting the result.

Integrated Throttle Control (ITC)

When a gas turbine–powered warship is under way, control of the propulsion gas turbine(s) is usually maintained either on the bridge or at a central (engineering) control station. Control of a gas turbine can also be maintained at a local operating station, although this is not the normal mode of operation for a twin-screw gas turbine–powered combatant.

On board *Spruance*- and *Perry*-class warships, control of the propulsion gas turbine(s) from the bridge is maintained at the ship control console (SCC). Control of both pitch and RPM of a shaft is exercised through an integrated throttle control (ITC) at the SCC. An ITC on board a *Spruance*-class destroyer is used to control the speed produced by its associated shaft in the following manner:

In the ahead direction the RPM of the shaft is maintained at 55 RPM, and ship speed increases as the ITC is moved to increase pitch ahead from the percentage which corresponds to zero thrust to 100% pitch ahead. At this point, further movement of the ITC will increase shaft RPM, and ship speed continues to increase as shaft RPM increases.

In the astern direction the RPM of the shaft is maintained at 55 RPM, and ship speed astern increases as the ITC is positioned so that the percentage of pitch which corresponds to zero thrust is moved to −49% pitch astern. At this point, further movement of the ITC will increase shaft RPM, and ship speed astern continues to increase as shaft RPM increases.

The ITC is positioned at a point that corresponds to the speed desired. The ITCs for both the port and starboard shafts may be mechanically latched together to control both shafts simultaneously. This is also true for the ITCs at the PACC (propulsion auxiliary control console) in CCS. Button action permits the EOOW to transfer control of both shafts to their ITCs in CCS or back to the bridge.

The single shaft of a *Perry*-class FFG is controlled by an ITC in much the same manner that an ITC is used to control its associated prime mover(s)/power train in a *Spruance*-class DD.

At this point you, as the prospective officer of the deck of a gas turbine–powered warship, are becoming sufficiently familiar with gas turbine propulsion to understand basically how the EOOW and the rest of the mobility team can get your ship under way using the Engineering Operational Sequencing System (EOSS).

A few pages from now, we'll go through a basic light-off sequence with the EOOW, our propulsion plant partner, whose general responsibilities were discussed in Chapter 5. Getting "under way" will acquaint you with normal procedures and terms such as "dead shaft pick-up" and "plant mode logic," with which the OOD of a *Spruance*-class DD must be generally familiar. A *Spruance*-class OOD's counterpart in another gas turbine–powered warship may find this section and these terms helpful in understanding how his own plant operates.

However, there are still a few items of general information about basic gas turbine propulsion that the Captain of any gas turbine–powered warship would want his officers of the deck to know.

For example, the Captain would want you to know that there are normal operating *parameters* of the propulsion gas turbines, including having an idea of what would happen if these parameters were exceeded. He'd like for you to have an understanding of what constitutes a *peril* to a gas turbine. To increase your confidence as an OOD, the Captain would insist that you have a general understanding of some features that *protect* gas turbines and the men who operate the propulsion plant itself.

Even if the Captain didn't ask, you'd probably want to know something about basic gas turbine *performance* (perhaps in case he did ask!). You'd also be interested in why the gas turbine–powered warship performs with such *quality* in terms of responsiveness, dependability, and efficiency/economy.

To use a familiar American phrase, "minding your 'P's' and 'Q's' " about your gas turbine propulsion plant can help make you become an OOD who is "comfortable" with engineering.

SOME "P's AND Q's" ABOUT GAS TURBINE PROPULSION

Parameters

Any prime mover or propulsion system has normal operating parameters, and those of gas turbines are closely monitored by both men and control systems. Control systems can be programmed to shut down a gas turbine automatically if normal operating parameters are exceeded.

Parameters monitored and the locations at which sensors are installed vary according to manufacturer. For example, in a Coast Guard *Hamilton*-class cutter, excessive exhaust gas temperatures (EGT) of an FT4 propulsion gas turbine may necessitate an engine "shutdown." But in an LM 2500, an excessive gas stream temperature *at the inlet* to the turbine that drives the load (the power turbine) can create an engine "shutdown" condition.

Table 8-1 shows normal operating parameters of the LM 2500 gas turbine which powers *Spruance*- and *Perry*-class warships. This listing has been provided to enable the OOD or prospective OOD to appreciate how his gas turbine control system monitors values—and these are only a few—in performing its control functions.

Gas turbines are precision prime movers that cannot be adequately controlled without the use of automatic controls. However, *men*, rather than devices, *control* the propulsion plant of a gas turbine–powered warship, just as they do in the case of a conventionally powered surface warship. If the commanding officer directs, use of a *battle override* switch, which is provided in many gas turbine control systems, will prevent automatic "shutdown" if normal operating parameters are exceeded (with the exception of the power/free turbine overspeed parameter).

Table 8-1

Name	Value	Abbreviation (The exotic-sounding term used by engineers)
Gas generator speed	4950–9950 RPM (idle)	N_1
Power turbine speed	1600-3600 RPM	N_2
Power turbine inlet temperature	1000–1499°F	$T_{5.4}$ or "T-I-T"
Lube oil pressure	16–80 psig	None
Lube oil temperature	Not greater than 340°F at scavenge outlet	None
Gas generator vibration	00–4 mils (Alarm at 4 mils, stop initiated above 7 mils)	GG vibes
Power turbine vibration	00–7 mils (Alarm at 7 mils, stop initiated above 10 mils)	PT vibes
Compressor discharge pressure	235–250 psig	CDP or P_{s3}

Perils to a Gas Turbine

Although the acronym FOD consists of only three letters, what it stands for poses a great threat to gas turbines: gas turbines cannot tolerate *foreign object damage.*

After any work in the inlet ducting—and at regular intervals specified by the PMS—the engineer officer and the leading engineman or gas turbine specialist inspect the inlet ducting and compressor inlet to ensure that no material is present which could be ingested by the gas generator. Marine gas turbines are also protected from foreign objects by screens near the bellmouth inlet to the compressor.

Ice is classified as a foreign object; the effects of its formation could be disastrous to a gas turbine. Gas turbines are susceptible to icing because when air enters the gas generator inlet, its velocity increases rapidly, resulting in a temperature drop. If the ambient temperature of the air is only a few degrees above freezing, the velocity increase/temperature decrease will result in icing, since a certain amount of moisture is always present in air.

As a general rule, ice-forming conditions exist whenever the relative humidity is greater than 70% and ambient temperature is 41°F or less. If the temperature is between freezing and 46°F, and rain or wet snow is present, conditions are such that large chunks of ice could be formed, which could do great damage to compressor vanes and blading.

Demister pads are installed in air intakes in *Spruance*- and *Perry*-class warships to screen out moisture and particulate matter of a relatively small size. All marine gas turbines can be protected from icing by using bleed air to physically heat either components of the gas generator itself or the air that enters the gas generator. The LM 2500 is protected from icing by heating the air that enters the compressor.

Bleed air to combat icing does not come free. The gas generator must increase speed to produce enough bleed air to protect itself. This results in an increase in fuel usage and/or a reduction in total power output of the gas turbine.

In *Spruance*-class destroyers alarms activate when icing conditions are present. Bleed air for anti-icing can be enabled at either the turbine local operating console or from CCS, and valves in the system cycle to keep the temperature of air at the compressor inlet above 38°F. If the temperature of the air at the compressor inlet drops to 36°F, another alarm appears, and "blow-in" doors in the air intakes activate to ensure that the compressor receives a sufficient air supply (although the moisture content will now be increased).

(The above paragraph illustrates still another way a gas turbine control system can be programmed in order to ensure that propulsion is maintained.)

Protection of the Gas Turbine

In the previous subsections concerning parameters and "perils" of a gas turbine, it became apparent that alarms and automatic features are widely used to afford a high degree of protection to both propulsion gas turbines and SSGTGs.

The modern marine gas turbine plant is an extremely safe plant. The degree of safety built into the propulsion plants of today's gas turbine warships gives both operators and officers of the deck tremendous confidence in these propulsion plants.

The following sequence controlled by logic circuitry is designed to combat a fire in a gas turbine module of a *Spruance*-class destroyer:

If a fire (which would be detected by sensors within the module) occurs:

Fuel oil shutdown valves activate.
Fuel oil trip valves located under the engine close.
Bleed air system valves close.
The base enclosure vent damper closes.
The module cooling fan stops.
Both audible and visible alarms activate.

At the end of a 20-second delay sequence a fixed CO_2 system activates, and the enclosure is filled with CO_2 in sufficient quantities to extinguish even very large fires. The module itself is designed to withstand temperatures of 2000°F for 15 minutes.

The preceding example is only one of many that could be given to demonstrate the high degree of safety provided to marine gas turbines—and the men who operate these precision prime movers.

Gas Turbine Performance

A marine gas turbine "likes" to do its work on a cold, dry day. The reason for this is that a mass of cold, dense air at the compressor inlet is already compressed, so to speak. The compressor will have to do less work on the air for a given amount of fuel consumed by the gas generator.

Marine gas turbines are sensitive to even small temperature changes and variations in the barometric pressure at sea level. In general, an *increase* in temperature will have a negative effect on engine performance. The fuel control of the gas generator senses changing conditions at the compressor inlet and automatically adjusts fuel flow and gas generator speed to produce the desired gas generator output.

The various manufactuers of gas turbines refer to a "standard day" as the basis on which to rate and evaluate their gas turbines. (Standard days, however, are not standard for all manufacturers.) Periodically the engineer officer or leading engineman/gas turbine specialist uses recorded data to determine if his gas turbine has lost efficiency when compared to its rated output on a standard day. As an example of overall gas turbine performance, an LM 2500 is designed to produce full efficiency on a day when the temperature at sea level is 100°F.

As the OOD or a prospective OOD, you may hear the engineer officer refer to the term standard day when discussing overall engine performance. The term "waterwashing" will probably be used often, too.

Because marine gas turbines operate in salt air, they must be cleaned frequently to prevent performance loss from salt or other contaminants which can form deposits on compressor vanes and blades. Waterwashing is accomplished by motoring the gas generator for a specified period of time and spraying distilled water and/or approved cleaning agents into the compressor. Waterwashing should be done after every engine shutdown.

Quality

When we speak of the propulsion plant quality of today's gas turbine–powered warships, the features that come most quickly to mind are its responsiveness, dependability, and efficiency/economy. The marine gas turbine propulsion plant rates extremely high in the first two categories. In the third category, that of efficiency/economy, claims by the opponents of gas turbine propulsion for warships that these ships would be "fuel-guzzlers" simply have not materialized.

The responsiveness of a marine gas turbine propulsion plant is evident in two areas. In terms of acceleration/decleration, *Spruance*-class destroyers are unmatched by steam turbine–driven destroyers, guided missile destroyers, and cruisers. *Perry*-class FFGs can outsprint FFGs 1–6 and the *Garcia*-class FFs, even though both classes of these ships are powered by the highly responsive 1200 PSI pressure-fired steam generating plants. And while the twin gas turbines of the Coast Guard's *Hamilton*-class cutters develop about 5000 total shaft horsepower *less* than the HP/LP steam turbines of a *Knox*-class FF—a warship roughly similar in size—a *Hamilton*-class high endurance cutter (WHEC) is faster than a Navy FF. (WHECs also have two diesel engines for propulsion, which are used at low and moderate speeds.)

The second area of responsiveness in which gas turbine–powered warships excel is their ability to have their prime movers available for propulsion in a very short time.

On board *Spruance*-class destroyers, for example, there are three methods for lighting off and engaging a propulsion gas turbine to the propulsion power train. These methods are as follow:

Manual control: The operator performs key functions with button action at a console. Various lights and data indications confirm that that action has been completed before the operator will use button action to perform another sequential event. Using EOSS and manual mode starting, an operator can have a gas turbine available for propulsion in about 20 minutes. (Close attention to detail is required, since this mode bypasses permissives of the start logic circuitry.)

Manual initiate: This mode of starting is available at either the local operating or the central control station. Button action by the operator initiates engine starting using start logic. After a designated key event has been completed, additional button action is required by the operator to continue the sequence of events. Separate button action is required to connect the prime mover to the power train after engine start. This method of starting may be thought of as "semi-automatic" operation. Using manual initiate, an operator can have a gas turbine available for propulsion in from two to four minutes.

Auto(matic) initiate (not available at the local operating station): Button action by the operator begins a sequence of events that culminates in engine start and power train engagement in a very short time period. As in manual initiate, the operator can stop auto initiate with button action at any point in the sequence of events.

A change in plant configuration—going from secure to split plant (one engine per shaft), from split plant to full power (both turbines clutched to the shaft), change engines, etc.—can be accomplished rapidly by proper use of the *start mode change* and *plant mode select* pushbuttons.

Propulsion plant controls for *Perry*-class FFGs are basically similar to those for *Spruance*-class DDs, although some electronic controls and logic circuits differ. In general, the controls of *all* gas turbine–powered warships provide a tremendous degree of responsiveness.

Dependability is achieved through reliability of control systems, responsiveness of those systems and the gas turbines themselves, and duplication of equipment. With the exception of a catastrophic failure of the electrical generation/distribution system, the *only* major threat to maintaining propulsion of a single gas turbine propulsion plant would involve a casualty to either a power train component or a component of the propulsion shafting associated with that plant.

With regard to efficiency and economy, gas turbine propulsion is not a particularly economical type of propulsion at low speeds. However, the combination of gas turbine propulsion and the CRP propeller enables gas turbine–powered warships to achieve the endurance requirements specified by military and naval planners. A "trail shaft" procedure, which has been developed for *Spruance*-class warships, indicates that twin-screw gas turbine ships may be capable of more overall economy than was previously anticipated. A summary of the trail shaft procedure follows:

In the trail shaft configuration, one gas turbine is used to drive its associated shaft. Both secured engines are declutched from the other shaft, which is permitted to "free wheel" with propeller pitch on that shaft set at 100% pitch ahead and the reduction gear lubricating and CRP oil systems operating.

Substantial fuel savings, particularly at low speeds, make this a Navy-approved method for transits and routine steaming under conditions of reduced readiness. The top speed that can be maintained in this configuration is comparable to that of a 1200 PSI CG operating with one boiler providing steam to both shafts at 95% boiler load (cross-connected operation).

At this point, we've reviewed some important "P's" and "Q's" of gas turbine propulsion. Now we can use this knowledge to understand how the EOOW prepares the ship to get under way. The example given in the following section will be for a routine light off under typical conditions.

The final section of this chapter will be devoted to types of gas turbine casualties and their effects on ship propulsion.

LIGHT OFF AND UNDER WAY IN A GAS TURBINE–POWERED WARSHIP

(This example is a typical sequence of events on board a *Spruance*-class destroyer, to be used for illustrative purposes only.)

Prior to directing any button action that would initiate engine starting and power train movement, the EOOW would ensure that each engine-room had the following systems aligned and in operation:

Fuel oil system
Lube oil system
Bleed air system
CRP system

Then—when you asked—the EOOW would tell you why *dead shaft pick-up* is required to get the ship under way: "Not long after *Spruance* destroyers began operating, we found that trying to clutch that first engine to its power train to get under way in the manner originally intended by designers was damaging numerous clutch/brake assemblies."

The EOOW would continue: "What was happening was that the clutches just couldn't withstand the forces needed to overcome the inertia of getting those reduction gears and its propulsion shaft moving. So the experts decided that by clutching the engine we wanted to start first to the power train and then starting that engine, we'd be able to overcome the force of inertia.

"Some people think dead shaft pick-up is a tricky maneuver because it can result in getting movement of the ship when you're still moored. But there's no problem if dead shaft pick-up is done correctly."

Engine start/power train movement using dead shaft pick-up can be accomplished from either the local operating station or the central control station. Even from CCS the EOOW would probably prefer to start the gas turbines in *manual initiate*, since it allows engineroom watchstanders an opportunity to check reduction gears and various control elements during the engine start sequence.

To start the entire sequence, the EOOW would ensure that throttle control is in manual. Pitch is in manual and set at 0%. Manual PLA for the engine to be started is set at 0%. (PLA stands for power level angle, and can be thought of as throttle control.) At this point the shaft brake is applied and the clutch engaged. Manual initiate is used (usually) to start the selected gas turbine.

During the sequence, logic releases the shaft brake. The (friction and dental) clutches remain engaged. When the turbine start sequence has been completed, the energy supplied to the power turbine by the gas generator at idle *may* be sufficient to overcome the inertia of the reduction gears and propeller shaft. If gas generator speed is insufficient, increasing the PLA will increase gas generator speed.

As soon as shaft rotation is apparent, conditions are stabilized to produce zero pitch thrust and 55 shaft RPM. At this point the throttle may be set for automatic operation. Automatic control of both pitch and shaft RPM may be exercised from either the bridge or CCS using the ITC.

After this procedure has been duplicated on the other propulsion plant, the ship is ready to get under way.

Suppose that shortly before getting under way, the Captain directs that full power be available for both shafts. To effect this change from split plant to full power, the EOOW orders the change to be made using *plant mode logic,* a method of accomplishing plant configuration changes that is characterized by reliability and simplicty.

To make the change to *full power,* the console operator in CCS ensures that he has control of all propulsion gas turbines. He then depresses the appropriate *start mode change* and *plant mode select* pushbuttons. Control logic will start the second engine for each shaft, clutch that gas turbine to the power train, and balance the load between both gas turbines now driving the shaft. This change can be made with virtually no fluctuations in shaft speed.

Now all lines are taken in, and the (*Spruance*) gas turbine–powered warship is under way.

The final section of this chapter will list casualties to the propulsion gas turbines and their associated power train/propulsion shafting that could affect the warship's maneuverability while under way.

PROPULSION PLANT CASUALTIES–GAS TURBINE–POWERED WARSHIP

The following casualties can cause loss of a propulsion gas turbine with an accompanying reduction in ship maneuverability:

- Gas generator overspeeding
- Power turbine overspeeding
- Propulsion turbine failure to reach idle speed (stall condition)
- High power turbine inlet gas temperature (stall condition)
- Low lube oil supply pressure
- Lube oil overheats
- Flameout
- Excessive power turbine vibration
- Excessive gas generator vibration
- Fire in propulsion turbine/module

The following casualties can result in loss of a propulsion plant for prolonged periods:

- Loss of CRP oil pressure (until emergency pitch can be locked)
- Loss of reduction gear lube oil pressure
- Roaring noise in the reduction gear
- Wiped line shaft (spring) bearing

If the tactical situation permits, it is recommended that the officer of the deck of a twin-screw warship slow to a speed on the unaffected shaft, as recommended by the engineer officer of the watch, that will expedite the evolution in any casualty involving stopping and locking a shaft.

CHAPTER REFERENCES

The Aircraft Gas Turbine Engine and Its Operation, Pratt and Whitney Aircraft Operating Instruction 200, May 1974 Revision

Aircraft Gas Turbine Engine Technology, Irwin E. Treager, McGraw-Hill Book Company, 1970

Gas Turbine Fundamentals, Pamphlet No. 139, U. S. Coast Guard Institute, May 1968

Jet Aircraft Power System, Jack V. Casamassa and Ralph D. Bent, Third Edition, McGraw-Hill Book Company, 1965

Principles of Naval Engineering, Chapter 23, NavPers 10788-B

Propulsion Plant Manual for SPRUANCE Class, NavSea 0941-LP-054-1010, 9 September 1976

9

Inspections, Efficiency, and Economy

If this book has met its objective, you are more comfortable with your general knowledge of the engineering plant on board your ship—or the ship to which you will be assigned—than you were before you started reading.

The engineering plant, including the propulsion system, of a modern warship is a tremendously reliable and responsive system that enables the ship to accomplish her mobility missions and to support all warfare missions. You, as the officer of the deck or a prospective OOD, the engineer officer of the watch—your propulsion plant partner—and the other engineers are all members of the mobility team responsible for proper operation of the plant.

Proper operation of the engineering plant is vital to protect the dependability of this responsive, efficient, and expensive installation. Thorough inspections and effective corrective action of deficiencies noted therein are also vital to protect the dependability of your engineering plant.

Some inspections, such as those conducted by the fleet commander's propulsion examining boards (PEB), are formal ones that examine actual plant operation, maintenance, cleanliness and preservation, and state of training of mobility team personnel. Another extremely formal inspection, by the Board of Inspection and Survey (INSURV board), is an inspection of the entire ship, required by Congress. Upon the conclusion of an "INSURV," the board is required to report its findings with regard to the ability of the ship to carry out her assigned wartime missions and whether the ship is fit for continued service.

Aside from the inspections noted above, the engineers inspect their plant as a matter of routine, both inport and under way. Periodically the Captain also conducts a "zone inspection" of the engineering spaces, as well as other departmental spaces throughout the ship. Eventually, you and your shipmate, "Ensign Benson," will find your names included in the list of those assigned to assist a more senior officer in his inspection of the engineering spaces.

Perhaps Ensign Benson is the electronic warfare officer or the assistant navigator. Perhaps Mr. Benson has been on board only a relatively short time. Perhaps all he can remember about a conventional propulsion plant is that the boiler "manufactures" steam and "sells" it to the main engine. Perhaps Ensign Benson is apprehensive about stepping into a main machinery space with equipment in operation, just as many of us were at one time.

At this point, Ensign Benson or any other non-engineering-oriented officer can take one of two approaches.

In the undesirable approach, Ensign Benson merely tags along during the zone inspection, wondering why the physical layout of main steam piping doesn't resemble the neat red lines he traced on clean white paper in an air-conditioned classroom. Since he didn't want to display his unfamiliarity with engineering, Mr. Benson didn't ask the senior zone inspector if that officer wanted Mr. Benson to concentrate on any particular area or equipment during the inspection. Mr. Benson will also refrain from asking questions during the inspection, since he really doesn't believe the engineers' adage that "there's no such thing as a stupid question asked by a man who really wants to know."

So, Ensign Benson wanders through the maze of pumps and piping, controllers and cable runs—basically wasting his time. He may discover a pocket or two of grime in an angle iron, but his inability to ask questions will prevent him from learning—or contributing—anything. By the time the inspection party finishes its assignment, Ensign Benson will not even have learned that steam lines must be looped and bent according to specification to allow for thermal expansion and contraction.

The alternate approach Ensign Benson may take to his zone inspection assignment is one used by numerous competent inspectors, including those with the benefit of many years of experience. If he chooses this approach, Mr. Benson will adopt the manner of an intelligent officer or petty officer who is unfamiliar with the space he is inspecting. However, this inspector has done some "homework" on the space. He knows what should be in it, and he is not afraid to ask questions.

Here's what will happen to Ensign Benson using his approach:

Mr. Benson has done some homework. Among other items, he knows that each main space should have a certain number of fire stations. The space should be clean so that equipment can be easily identified and no hazards to men and machinery exist. When he enters the space, Mr. Benson begins to ask himself and others simple questions. The answers in most cases will be immediately evident. For example:

Is each fire station made up correctly? The All Hands Damage Control Personnel Qualification Standard (PQS) for the ship indicates what firefighting equipment should be at each station. Is any gear missing or inoperative?

How could a fire start? Are flanged connections in fuel and lube oil piping protected by effective flange safety shielding, or does that pool of oil underneath mean that oil could escape under pressure and spray onto hot, unlagged steam piping or an unshielded light fixture nearby? Could a fire begin by water leaking from a heat exchanger, cooling line, or valve stem drip or spray into a nearby electric motor or power panel, shorting out that motor or power panel? Is the power panel completely enclosed? Are incandescent lights shielded? Is lagging in place? Does the lagging appear to be in reasonably good condition, or are areas exposed, or is the lagging oil-soaked?

What would happen if a fire did start? Could the fire feed on pockets of oil in the bilges or the oil leaking from a piece of machinery which seems to indicate that the equipment needs repair? Could the fire feed on the dust inside and travel along the ventilation duct that looks dusty when Ensign Benson shines his flashlight into the ducting? Are watchstanders generally familiar with their firefighting equipment?

What would happen if the fire resulted in an explosion, and the space had to be evacuated? Are deckplates secured with at least two fasteners set diagonally so that the deckplates will not become missile hazards, threatening the life and limbs of men as flying deckplates did during the fireroom explosion on board a U. S. DDG moored in Toulon, France, in the late 1960s? Are ladders and handrails leading to the escape trunk securely in place? Does the escape trunk door provide a fire-tight seal? Is the escape trunk lighted and free of obstacles, or will its use as a storage area make the trunk a death trap for men as they stumble into it? Does the space evacuation alarm at the top of the escape trunk ladder activate? And were all remote cables connected to valves below so that damage control personnel could secure the space from topside or another remote location?

Simple questions, certainly. Still, Ensign Benson could not have asked any that are more important, because these deal with the safety of men. At this point Mr. Benson still does not know whether a "guarding" valve is the same as a "guardian" valve, nor has he begun to look for evidence of corrosion and deterioration caused by the use of dissimilar metals. But more about these subjects later. For now, Ensign Benson has made a good start.

PREPARING TO INSPECT

Engineering material inspections fall into two basic categories: *space* inspections and *operational* inspections. Some inspections, such as an operational propulsion plant examination (OPPE) conducted by a fleet commander's PEB, or an underway material inspection (UMI) conducted by the INSURV board, are both operational and space inspections.

Until he is assigned to the engineering department, Ensign Benson will not become involved with operational engineering inspections. Even if Mr. Benson eventually becomes the main propulsion assistant, or the boilers officer in a larger ship, most of the engineering inspections in which he participates will still be space inspections.

Preparing for a space inspection such as the Captain's zone inspection of the engineering department isn't difficult. Ensign Benson has already made a good start by concentrating on safety, which is a primary goal of any inspection. Mr. Benson may also find that his homework on engineering "hardware" presents no particular problem. Chapters 3 and 4 of this book, for example, discuss what equipments are found in the firerooms and enginerooms, respectively, of 1200 PSI CGs, DD/DDGs, and FF/FFGs. The Propulsion Operating Guide (POG) of all naval ships indicates where equipment is located within a particular engineering space, listing such equipments as air compressors, fresh water pumps, and even antisubmarine rocket (ASROC) heating and cooling pumps (in applicable ships), along with propulsion equipment. Even if Mr. Benson temporarily forgets his homework, he still has little reason for concern, since operating instructions posted near each piece of equipment can help him distinguish a feed booster pump from a fuel oil service pump or a pump used for another purpose.

Actually, Ensign Benson has already prepared himself to be a competent inspector if he is unafraid to ask questions. Here's just one example:

While inspecting near the main feed pumps, Ensign Benson discovers that the immediate area near the turbine end of one main feed pump is much hotter than the corresponding area of another main feed pump. Before this, Mr. Benson had been concentrating on the condition of lagging and checking the base of the pumps to see if any lube oil was present. But now he becomes interested in the readily apparent difference in heat.

In reply to Mr. Benson's question, the fireroom supervisor gives this report: "It's hotter around that particular pump because the steam seals need replacing. Steam seals mechanically prevent steam from escaping from the turbine ends, and it's a major job to replace the steam seals. Yes, we do have that deficiency listed on the CSMP [Current Ship's Maintenance Project, a document maintained as part of the Navy's Planned Maintenance System]."

Or, the fireroom supervisor might give this report: "The steam seals on that main feed pump need replacing, but the real problem is the auxiliary gland leak-off system. The vacuum in the gland leak-off condenser is so low that I think the condenser salt water side needs cleaning."

When Ensign Benson asks how the auxiliary gland leak-off system functions, the senior petty officer replies: "It's a system on our ships and other newer warships that takes low pressure steam vapors from the turbine glands on larger pieces of equipments and condenses these vapors for return to the feed system. On our particular ship the fan on the gland leak-off condenser also removes air from the deaerating feed tank.

"Actually, we could operate without the auxiliary gland leak-off system for a short time if we had to. But that would mean that we'd have to use a great deal more makeup feedwater, and this space would get so hot and humid that the heat would be unbearable."

Since Mr. Benson is unafraid to ask questions, it's obvious that he can learn a great deal from knowledgeable petty officers during a space inspection as well as make a contribution to the inspection by noting deficiencies and problems.

Actually, Ensign Benson or any other young officer can become a competent, contributing inspector of the engineering plant (or any work center) by observing some basic inspection techniques. These include the following:

• Ensure that safety equipment is maintained properly. Nothing is more important than the life of a shipmate.

• Ask questions. It's possible to learn a great deal during inspections because your shipmate will be responsive if he knows you take an interest in his work.

• Don't worry about discovering all problems in a particular space. Space inspections in particular require repetition. A logical approach taken by a new inspector may uncover some problem so obvious that even more experienced personnel have overlooked it.

• Look for major items. One unsecured deckplate may seem insignificant, but an explosion or the near miss of a missile impacting close aboard could turn that deckplate into a lethal weapon inside the ship.

• Ensure that deficiencies are recorded so that corrective action can be taken. The Navy's PMS requires that a space work log be maintained. If the space supervisor says a particular deficiency is already logged, his formal work list or CSMP should indicate this.

• Remember that as an officer of the deck or a prospective OOD, you are a key member of the ship's mobility team. The object of an engineering space inspection is to protect the safety of shipmates and the reliability of the plant, not to develop a list of insignificant "I-gotcha's."

ENGINEERING TERMINOLOGY (OR ENSIGN BENSON VS. VALVES)

Before a further discussion of engineering material inspections and the *goals* of inspections—and on a lighter note—one word of warning may be helpful to Ensign Benson or any other non-engineering-oriented officer about to embark on an inspection of an engineering space: Engineering terminology can sometimes be tricky. Take the subject of valves, for instance:

If Mr. Benson asks, the engineers will tell him that stop valves include globe valves, gate valves, plug valves, piston valves, needle valves, and butterfly valves.

At the exact moment that he remembers that *stop valves* are used to *prevent* or *restrict flow,* the engineers begin to tell Mr. Benson about the other basic valve type, the *check valve.* Check valves include ball-check valves, swing-check valves, and lift-check valves, which all act to *permit flow of a fluid in only one direction.*

The engineers add that numerous valves in the engineering plant are stop-check valves, which function as either, depending on the position of the valve stem. Then they open a floodgate of other types of valves: relief valves, unloading valves, sentinel valves, reducing valves, and diaphragm-controlled valves. These are types of *special purpose valves,* and this list includes a type previously encountered—boiler safety valves, which the OOD knows must lift in sequence to protect the boiler from overpressures.

The engineers may want to continue by talking about special applications of stop valves such as bypass valves, which are used to equalize pressures and temperatures around larger valves, and those highly important boiler blowdown valves.

If Ensign Benson has been unable to take much time from his primary duty for studying valves, he is suddenly aware that there are many different kinds of valves. If he doesn't remember that a valve is basically *a device for preventing or controlling the flow of a fluid through an opening,* he may have a difficult time with engineering terminology in his first few zone inspections.

Guarding valves, in particular, may give Ensign Benson some grief. This can occur, not as the result of a particular guarding valve leaking or being difficult to locate, but as a result of imprecise use of the term by some engineers. Then there's another group of engineers who have grown up calling a "guarding valve" a "guardian valve," and if Mr. Benson is struggling to become comfortable with engineering, he may suddenly become uneasy over the suspicion that he now has still another type of valve to consider. So, let's discuss guarding valves with Mr. Benson:

In the case of main engine guarding valves, General Specifications (GenSpecs) for Ships 1973 prescribes the exact location in both multi- and single-plant ships. In multi-plant warships, a main engine guarding valve is located immediately upstream of the main steam strainer (which is located upstream of the ahead and astern throttles) for each propulsion unit. In single-plant warships a guarding valve is located immediately upstream of the throttle valve to the HP turbine (ahead throttle), and another guarding valve is located immediately upstream of the throttle valve to the astern elements of the LP turbine (astern throttle). If a main engine throttle casualty occurs, the engineers can use the main engine guarding valve to secure steam to the turbines or as a throttle valve itself, as the situation requires.

Engineers who use the term "guarding valve" in reference to the stop valve upstream of the throttle valve of an auxiliary turbine are actually referring to the "root (or root steam) valve" for that particular pump. It should be noted that GenSpecs 1973 also requires that a second stop valve be installed as a cut out valve between the root valve and the throttle/control valve *when* the throttle/control valve is integral to the equipment, which is often the case with pumps such as main feed pumps and turbine-driven fuel oil service pumps.

Guarding valves are also important protective valves in boiler blowdown systems. Figure 9-1 shows the location of guarding valves, which provide two-valve protection for a boiler when it is off the line and also protect the bottom blow piping from becoming pressurized in case header and drum blowdown valves leak through on a steaming boiler.

But so much for the subject of valves, which are adequately explained in numerous engineering texts and technical publications, including *Principles of Naval Engineering* and the applicable chapters of NavShips *Technical Manual,* respectively. (Technical manuals for specific valves on board ship are maintained in the engineering technical library, which provides still another source of information about valves for the engineers on board ship.) Ensign Benson is now sufficiently familiar with the names of valves to ask questions during zone inspections. He knows that engineers who speak of "guardian valves" and "guarding valves" are talking about valves that perform the same function.

Let's assume that Ensign Benson has done enough engineering homework to know what basic equipments he will find in each engineering space on his zone inspection route. He is unafraid to ask questions, even if this requires asking the space supervisor to clarify terminology with which Ensign Benson is unfamiliar. With this initiative, desire to learn still more, and his "fresh" set of eyes, Ensign Benson will find that he can make a definite contribution during engineering material inspections.

ENGINEERING MATERIAL INSPECTIONS

As mentioned previously, engineering material inspections are either *space* inspections (a "zone" inspection is a type of space inspection) or

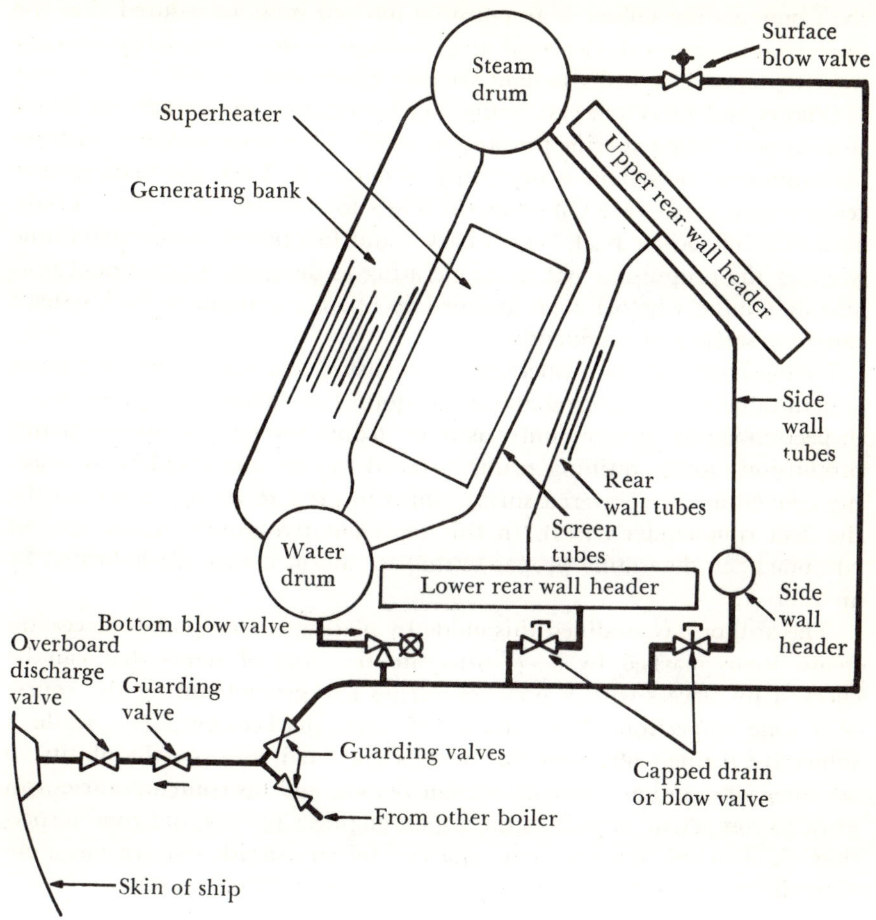

Figure 9–1. Simplified boiler blowdown system
(for a single furnace, uncontrolled superheat boiler)

operational inspections or a combination of both. Line officers from other departments will find themselves assisting in space inspections of the engineering department, just as the line officers assigned to the engineering department will assist in space inspections of other departments.

The reason why inspections are so important to the engineering plant, as well as the entire ship, is best summed up in a memorandum from the president of the Navy Board of Inspection and Survey to prospective commanding officers at the Surface Warfare Officers School, Newport, R.I., in August 1972: "Today's Navy has evolved from the point where rope and gun barrel wear determine whether a ship can fight, to a point where the Commanding Officer is responsible for and must be assured that the numerous mechanical, electrical and electronic systems and subsystems are working together in order to fight the ship effectively."

Officers and petty officers conducting operational inspections on board ship inspect with a specific purpose in mind. They want to know whether an equipment or system is operating as designed. They use maintenance requirement cards (MRCs) from the PMS to conduct their tests. Trials, such as a full power trial, "crash back," and full power astern, determine whether the propulsion system is operating as designed. Operational tests also determine whether men assigned to operate equipment and systems can do so safely and confidently.

Engineering material inspections help ensure that the engineering plant is maintained in a safe, reliable condition. The following representative inspection guide is commonly used by (conventional propulsion plant) propulsion mobile training teams assigned to type commanders in assisting conventionally powered surface ships to prepare for an inspection by the fleet commander's PEB. In this representative guide, actual tests of equipment to determine proper settings or alarm points are indicated by an asterisk.

The author has modified this guide by placing crosses (†) beside certain items. Items marked by a (†) illustrate the type of items that can be checked by officers with a nonengineering background during the course of a zone inspection. The number of items marked by a (†)—while a subjective listing—indicates that with a good pair of eyes, the ability to ask questions, and a flashlight, Ensign Benson and his contemporaries can assist as competent inspectors during an engineering material zone inspection. (Unmarked items or items marked by an asterisk require more engineering expertise to determine correct setting and conditions.)

FIREROOM

BOILERS

SAFETY VALVES

† Harness padlocked

Lead wire and seal on blowback cap

† Short studded safety valves studs (Check with MPA for a definition of short-studded connections on lines of 150 psi or greater.)

Hand easing gear properly adjusted

† Lifting nut clearance

BURNER FRONT

* Burner return check valve under pressure for leaks

* Burner safety cut out valve under pressure for leaks

BOTTOM BLOW VALVES

† Locking devices

† Warning signs installed
Adequate operating gear available

GAUGE GLASSES

† Clean
† Leaks
† Chain guard on glass cut outs
† Normal water level indicated
* Gauge level agrees with remote indicators

FUEL OIL SYSTEM

* FO service tank QCV
* FO strainers interstage leak, shields, drains
† FO flange shields
* Recirculate fuel oil (leaks)

ALARMS

* Low water
* High water
* Salinity alarms
* Flame failure alarm
* In line desuperheater alarm
* All other IC alarm panels energized

DFT

* DFT high water alarm
* DFT low water alarm
† Thermometer calibration
† Brazed joints on DFT sample cooler supply lines
† DFT sampler station operating instructions installed

SOOT BLOWERS

† Operating instructions installed
† Soot blower drain valve warning signs
Freeness of scavenging air valves
† Chain guards on operating gear wheels
* Operation of elements (hot and cold)

REMOTE CABLES (operable)

†* Main stop (covers over remote hand wheels)
(Remote indicator lights)
†* Auxiliary stop (covers over remote hand wheels)
(Remote indicator lights)
* Fuel oil return valve (cover over remote hand wheels/pulls)
* Feed check valve
* FDB local controls

BOILER MISCELLANEOUS

Superheater backfill line cap $\frac{1}{16}$ inch drilled hole

† Chemical injection tank lagged and screen installed in tank
† Operating instructions installed for:
Chemical injection tank
Boiler sampling stations
Bottom blow operation
Surface blow operation
Boiler purge table posted at boiler front
* Cold and hot ABC inspection
† Lagging and lagging pads on boiler and steam lines
Shields over capillary tubing to remote boiler thermometers

MAIN FEED PUMPS

* Low lube oil alarm
* Low suction trip
* Hand trip
* Automatic start of electric lube oil pump
* Hand-operated lube oil pump output
Discharge relief valve sign
Locking device on recirc valves
Lube oil strainer shield and inspect internals
† Lube oil flange shield
† Lube oil leaks
† Lagging and lagging pads
Lube oil sump drain valve and plug
† Operating instructions installed
Lube oil strainer vent lines welded
† Lube oil samples
† Coupling guard installed

MAIN FEED BOOSTER PUMPS

* Low suction trip alarm
* Low discharge pressure alarm
* Automatic start feature
Locking devices on recirc valves
† Lagging pads on pump end
† Lube oil flange shields
Lube oil sump drain valve and plug
† Operating instructions installed
Inspection covers wire-sealed

FORCED DRAFT BLOWERS

* Overspeed trip
Lube oil strainer shield (inspect internals)
Lube oil sump drain valve and plug
† Lube oil flange shields
† Lube oil leaks

† Lagging and lagging pads
Inspection covers wire-sealed
Damper locking devices
† Lube oil samples

AIR COMPRESSORS

* Low ACC air pressure alarm
† Lube oil and/or salt water leaks
† Operating instructions installed
† Pulley belts worn
† Pulley belts protective screen
Inspection covers wire-sealed
† Lube oil samples
* High temperature cut out switch

FIRE STATIONS

AFFF hose kinked or cracked—operating instructions
† All CO_2 and PKP currently inspected
† All hoses currently hydrostatically tested
† Fixed CO_2 currently inspected
† Shield over 50-lb CO_2 bottles

CONTROLLING INSTRUMENTATION CALIBRATION (on main gauge board)

† Boiler drum psi (Calibration sticker indicates recalibration is not yet required.)
† Superheater outlet psi
* Superheater outlet temp (direct and indirect)
* Desuperheater outlet temp (direct and indirect)
Desuperheater outlet psi
† Fuel oil service main psi (calibration sticker)
FOSP discharge psi and temp
Fuel oil return psi and temp
Auxiliary exhaust inlet to DFT temp
† Auxiliary exhaust psi
* MFP discharge psi
* 600 and 150 psi
* FDB tachometers
* Air box wind indicators
* Remote and local boiler water level indicators
* Remote and local DFT indicators

LOCKED OPEN OR CLOSED VALVES (LD—locking device/L—padlock/WS—wire seal)

Boiler safety valves harness (L)
Boiler blowback plug (WS or L)
† Line shaft bearings (LD or L or WS)

MFP lube oil rundown valve (closed LD)
Ballasting/deballasting valves to salt water (closed LD)
Feedwater tanks (LD)
Bottom blow valves (LD)

ESCAPE TRUNKS

† Adequate emergency and normal lighting
† Flame-proof door operable
† Security of space (locked box on scuttle)
† Stowage in trunk
†* Evacuation alarm
† Ladders and trunk constructed of steel vice aluminum
† Ladders secured

FIREROOM MISCELLANEOUS

Sign on capillary tubing "Don't remove under pressure"
† Sign on line valve from superheater to desuperheater "This valve to remain open when boiler is steaming"
Relief valves (adjusted, tags, lifting gear)
Combination exhaust and relief valve warning signs
† Inventory of repair V
* Operation auxiliary gland leak-off system operation
† Deckplates secured (minimum of two fasteners secured diagonally)
† Short-studded connections on 150 psi and above lines (Check with MPA for number of threads that should be showing above top of nut.)

ENGINEROOM

MAIN ENGINE

* Low lube oil alarm
* Wrong direction alarm
* Loss of control air alarm
† Operating instructions installed
† Lagging and lagging pads
† Inspection covers wire sealed
† Lube oil samples

REDUCTION GEARS

† High security padlocks installed
† Inspection covers wire-sealed
† Main engine lube oil strainer covers (adequate and internals)
Welded lube oil line on LO strainer vents and HP valves
† Lube oil flange shields
† Oil leaks
† Thermometers calibrated
† LO samples

CONDENSATE PUMPS

† Oil leaks, steam and water leaks
† Lube oil flanges
† Operating instructions installed
Lube oil sump drain valve and plug
* Govenor operable
† Gauges and thermometer
Inspection covers wire-sealed
† LO samples

MAIN LUBE OIL SERVICE PUMP

† Lube oil leaks
† Lube oil flanges
† Operating instructions installed
Lube oil sump drain valve and plug
LOSP discharge valve—locking device—open
Relief valves and preservation
Inspection covers wire-sealed
† LO samples

MAIN CIRCULATING PUMP

† Lube oil leaks
† Lube oil flange shields
Salt water packing gland
† Operating instructions installed
LO supply drain valve and plug
Inspection covers wire-sealed
† LO samples

LUBE OIL PURIFIER

† Lube oil leaks
† Lube oil flanges
† Operating instructions installed
† Lube oil settling tank rundown valve—locking device—closed

EVAPORATORS

* High salinity alarms
† Salt water or evaporator distillate leaks
† Gauges and thermometers calibrated
† Operating instructions installed
† Lagging and lagging pads installed

SPRING BEARINGS

† Inspection covers—locking device or padlock—closed
† Lube oil drain valve and plug
† LO samples
† Thermometer installed and calibrated

SHIP'S SERVICE TURBO GENERATOR

- * Low lube oil alarm
- * High temperature alarm
- * Emergency trip
- * Overspeed trip
- † Welded lube oil vent line on strainer and HP valves
- † Lube oil flanges
- * Lube oil strainer shields (adequate and inspect interior)
- † Oil leaks
- Lube oil sump drain valve and plug
- † Operating instructions
- †* Tachometers calibrated
- † Thermometers calibrated
- † Security of covers and inspection ports
- † Lube oil samples

FIRE STATIONS

- † AFFF hose kinked or cracked—operating instructions
- † All CO_2 and PKP correctly inspected
- † All fire hoses currently hydrostatically tested
- † Fixed CO_2 currently inspected

ESCAPE TRUNK

- † Adequate emergency lighting
- † Unauthorized stowage in trunk
- † Flame-proof door operable
- † Security of space (locked box on scuttle)
- †* Evacuation alarm
- † Steel ladders and trunks

OTHER ENGINEERING SPACES

EMERGENCY DIESEL

- † Lube oil and fuel oil flanges
- † Lube oil and fuel oil strainer shields
- † Operating instructions
- * Overspeed trip correctly set
- Lube oil drain valve and plug
- † Reduction gear security
- * High temperature alarm
- * Low lube oil alarm
- * CO_2 flooding alarm
- * Flooding alarm
- Controlling instruments calibrated
- * Auto start capability

HP air leaks

† LO samples

Note: List for GTG and GTMs in gas turbine–powered ships will be similar, since GTG/GTMs are encapsulated in modules, which are to be entered only by authorized personnel.

CONTROLLING INSTRUMENTATION CALIBRATION

† All gauges on main throttle board calibrated

Temperature on air ejectors

* Flooding alarm

* Potable water alarm

* Evacuation alarm

* High salinity cell alarms

REMOTE OPERATED CABLES

†* Throttles

†* Bilge eductor valves

†* Bulkhead stops

AFTER STEERING

† Stowage in after steering and ram room

* Lighting

† Lube oil leaks

† Lube oil flange shields

MISCELLANEOUS

† Oil in bilges

† Ventilation ducts dirty—fire hazards

† Fire/missile hazards

Relief valves (adjusted, tags, and lifting gear)

† Salt water pump corrosion and packing glands

† Deckplates secured

CONTROLLERS/STARTERS

† Dirt/corrosion (exterior and interior)

† Operating instructions

MOTORS/MG SETS

† Dirt/corrosion (exterior and interior)

† Grounding straps installed—metal to metal contact between grounding strap and equipment/base foundations

† Operating instructions

† Vent screens clean

MAIN SWITCHBOARD

† Dirty/corrosion (exterior and interior)

† Instruments in calibration

† Rubber matting

† Grounds present (lights will indicate)
† Repair kit/first aid kit

**LIGHTING* (normal and emergency)

† Adequate
† Lighting covers installed and clean
† Battle lanterns operable

BATTERY LOCKER

† Clean
† "No smoking" signs
† Operating instructions
† Soda eye wash
† Rubber apron

**ABTs*

Operable
Clean

ELECTRICAL DISTRIBUTION PANELS

† Clean
Operable

PORTABLE ELECTRIC TOOL ISSUE ROOM

† Inventory of tools and personnel electric equipment
† Issuance of gloves, goggles with electric tools
† Records of inspection of tools
Knowledge of safety check procedures

ENGINEERING INSPECTION GOALS

Engineering inspections determine if the engineers and their engineering plant can perform required mobility and support missions. The primary goal of an engineering inspection is to ensure the safety of men and equipment. Important secondary goals are to ensure the efficiency—and reliability—of the plant, as well as its economical operation. The following sections of this chapter will briefly discuss safety, efficiency, and economy.

Within the engineering plant, the relationship between safety and efficient, economical operation is often a close one. The following example is only one of many that illustrate how safety can be related to efficiency and economy:

Earlier in this chapter, your shipmate, Ensign Benson, became acquainted with the auxiliary gland leak-off system. Malfunction of this system—which collects low pressure steam vapor from the turbine glands of main feed pumps, forced draft blowers, and turbine-driven fire pumps (as well as the DFT vapor condenser and/or fresh water drain collecting tank, in some ships) for return to the feed system—can lead to excessive loss of feedwater and high humidity within the space. (In some cases malfunctioning of this system has also caused bearing failure by heating lubricating oil in the equipment beyond its breakdown point, or allowing condensing vapors eventually to contaminate the lube oil at the turbine ends.) Periodically fireroom maintenance men must clean the salt water side of the gland leak-off condenser to ensure that adequate condenser vacuum can be maintained to collect all the low pressure steam vapors from the equipment the system serves. If the system is not operating properly, men will suffer from excessive heat and humidity, and more fuel must be burned to heat the relatively cool makeup feedwater that is required to compensate for the excessive loss of feedwater.

Now, suppose the fireroom maintenance crew has cleaned the auxiliary gland leak-off condenser salt water tubes and header, and reassembly appears to be all that is necessary to restore the system to use. Then someone discovers that some of the nonferrous studs used to connect the flanged headers to the condenser shell have been misplaced. In his haste to get the system back in operation, the petty officer in charge decides to substitute steel bolts for the missing studs. Dissimilar metals are now being used in this salt water heat exchanger, a condition that is sometimes encountered in salt water systems and lines throughout the ship.

Within a few months rust and corrosion begin to appear at both ends of the condenser. Supervisors are aware of the problem, but when the first cold iron period comes there is only enough time available to correct more pressing deficiencies. Zone inspections are conducted, but Ensign Benson and more senior inspectors fail to note the salt water crystals building up around the deteriorated steel studs. Soon salt water begins to drip onto equipment below. Then the drip becomes a steady leak, finally a stream of salt water. During heavy seas the ship begins to roll, and some of the salt water enters a power distribution panel mounted below.

Trouble?

Certainly, sooner or later. Perhaps the watch will be nearby and take effective action. Perhaps he won't. Perhaps the incident will occur during Ensign Benson's watch as junior officer of the deck. Perhaps the problem will be only minor. Perhaps it won't.

INSPECTION GOAL: SAFETY

The primary goal of any inspection is to ensure safety. The Navy spends many millions of dollars to procure equipment that is safe to operate. Additional millions are spent to develop safe operating procedures. Even more money is spent to ensure that the propulsion plant and auxiliary engineering spaces are environmentally safe.

Keeping the plant safe to operate involves additional expenditures of time, manpower, and money. But the product justifies the expense. The engineers will willingly steam their plant if they know they are operating under safe conditions. Use of that formidable list of alarms, gauges, warning plates, safety and shut-off valves, flange safety shielding, and numerous other items from the preceding space inspection guide helps the engineers ensure that they are operating their plant safely.

Engineers on board naval ships operate their plant according to procedures set forth in the Engineering Operational Sequencing System (EOSS). EOSS is composed of Engineering Operational Procedures (EOP), which are used for plant light off and securing, and Engineering Operational Casualty Control (EOCC). EOSS stresses safety, and the standardization provided by EOSS and PQS, the Personnel Qualification Standard (for individual watch stations), has done much to make members of the mobility team safety-conscious concerning the operation of their enginering plant. Still, an effective program of engineering inspections and corrective action is vital to the goal of safety.

Another program that enhances safety on board ship is the ship's formal tag-out program. Tag-out procedures apply to all systems—steam, electrical and electronic, and fluid systems. These procedures should be understood by all hands and especially by the engineers, since the engineers provide so many vital services as well as propulsion for the entire ship. Tag-out programs are now so important that the crew's knowledge of tag-out procedures is tested during PEB and other engineering inspections.

No program, however, can be effective without adherence to procedures. This depends on people who understand the importance of following procedures for the safety of themselves and others. Before we "sequence" to the subjects of efficiency and economy, the following episode is offered to illustrate the importance of formal programs such as tag-out procedures and electrical safety:

A former shipmate—let's call him "Ensign Benson," since he's still on active duty, and he says that this wasn't the happiest day of his life—was anti-submarine warfare officer on board the USS *McDermut* (a *Fletcher*-class destroyer) during the early 1960s. Since Mr. Benson was the ASW officer, he was concerned with removal of the 21″ steam torpedo tubes when the message came from the type commander directing removal of the "fish."

The job was scheduled for a Thursday morning, and a bit of nostalgia surrounded the whole affair: the *McDermut* was one of two destroyers credited with the sinking of the Japanese battleship *Fuso* in a night torpedo engagement during the Battle of Leyte Gulf. Nevertheless, it was time for the old steam torpedoes and the torpedo tube mount to go. Ensign Benson had a special reason for hoping the evolution would be concluded by Thursday afternoon: he was to be married in Los Angeles the following Saturday morning.

Work associated with removal of the torpedoes and mount appeared to be proceeding smoothly. First the "fish" were off-loaded, then the mount was unbolted from its base ring on the 01 level amidships. The electrician's mates reported that all electrical connections to the mount had been disconnected. But Ensign Benson went down to the main switchboard from which the mount was energized, just to make sure.

The leading torpedoman—the man in charge of stowage of the alcohol used in the torpedo propulsion motors—wasn't waiting, however.

Ensign Benson raced back to the maindeck. He arrived just in time to see his torpedoman, standing on a metal chair on the metal deck, attack a power cable with a pair of metal bolt cutters—

ZOT!!!

The petty officer's curly hair suddenly stood straight up. He landed against a bulkhead 18 feet away. The interior of the ship was plunged into total darkness. And suddenly the Captain appeared.

First the Captain assured himself that the torpedoman, although shaken, was breathing normally. Then the Old Man trained his sights on his ASW officer:

"BENSON-N-NN!! I don't care if your old man is an admiral! I don't care if you've invited the fleet commander to your wedding! I don't care if the president himself is going! All I know is that you're not going anyplace until all this wiring is repaired!"

But that wasn't the worst. The Captain roared. He ranted. He raved. Finally he concluded by promising Ensign Benson that those silver (lieutenant) jaygee bars, while oh-so-near, were in reality quite far away. About as far away, in fact, as the circle of stars a fleet admiral wears. Then the Old Man stalked off.

Sometime later, as Ensign Benson stood gazing at the Coronado shoreline, considering life in general and his choice of careers in particular, the torpedoman approached:

"Heck, Mr. Benson, don't take it so hard," his petty officer consoled. "We can have everything fixed up by next Tuesday easy."

Still, the leading torpedoman sensed that his words of condolence weren't doing quite enough to cheer up his division officer. So he gave Mr. Benson a tremendous smile and added: "Anyway, Mr. Benson, I'll bet you're glad that I remembered to wear these rubber gloves. I might not be breathing if I hadn't remembered to take safety precautions!"

(Before discussing efficiency and economy, a positive note: The casualty was restored by Friday, Ensign Benson was married on Saturday, and the Navy is experiencing fewer personnel casualties now than in past years, partially as a result of effective tag-out and electrical safety programs. Now, on to efficiency and economy.)

INSPECTION GOALS: EFFICIENCY AND ECONOMY

Efficiency and economy are closely related. Experienced engineers add that efficient, economical plants are also reliable and safe to operate.

The energy balance diagram from Chapter 1, repeated here as Figure 9-2, shows that some energy losses are unavoidable. However, some other energy losses—such as boiler air casing leaks, excessive gland leak-off, and many steam leaks—can be prevented or at least reduced. Inspections identify these energy losses.

These three common heat losses and their effects may be described as follows:

- *Boiler air casing leaks* can be detected when the forced draft blowers are supplying air for combustion. Many can be spotted visually by discoloration of the outer casing surface. Numerous casing leaks require that forced draft blowers run at correspondingly higher speeds to provide the required amount of air for proper combustion. This in turn means that greater amounts of motive energy must be supplied to the forced draft blowers. In addition, casing leaks raise the ambient temperature of a fireroom or main machinery room. The alertness of watchstanders decreases as ambient temperature increases.
- *Excessive gland leak-off* has been discussed earlier in this chapter. In addition to the overall loss of feedwater, both heat and humidity increase within an engineering space. Both will affect the alertness of watchstanders.

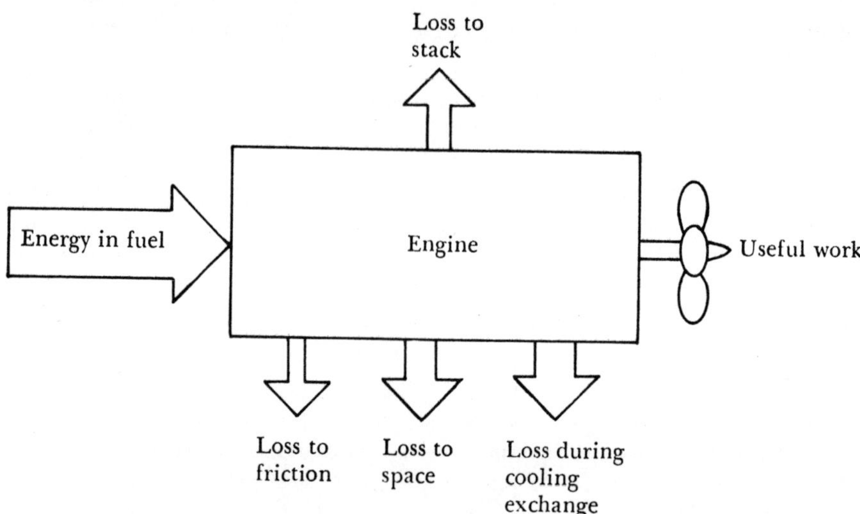

Figure 9–2. The energy balance

• *Steam leaks* occur whenever maintenance has been ineffective or neglected. Steam leaks represent another loss of steam, and therefore heat and fuel losses. Steam leaks raise both heat and humidity within a space, thereby affecting watchstanders. Moreover, a serious steam leak can be dangerous—a broomstick used to investigate whenever a loud whistling noise associated with a serious steam leak is heard can be chopped into pieces by the force of the escaping steam!

The cumulative effect of boiler air casing leaks, excessive leak-off, and even a few steam leaks can prevent a conventionally powered ship from making its required full power. These common material problems can be identified through a program of thorough and routine inspections.

For many, the subjects of efficiency and economy relate only to fuel usage. Fuel usage is the most important determinant of efficiency and economy. But efficiency and economy can also have a direct relationship to equipment usage and the men who operate the plant. The practice of steaming "modified main," referred to in an earlier chapter, affords a good example of the relationship between engineering inspections, efficiency, economy, increased equipment usage, watchstanding requirements, and correction of the deficiencies noted on inspections.

Conventionally powered surface warships routinely steam modified main before getting under way from the pier or anchorage; as the term implies, the main engine is in operation. But compared to steaming with just turbogenerators and their associated auxiliary plants, steaming modified main is not an economical practice. Let's take a typical situation:

Two DDGs are nested together during a working day inport. Shore power is unavailable, so both ships are steaming. Each must use two of its four 500 KW SSTGs, but one DDG must steam modified main. Moreover, the ship steaming modified main is having difficulty holding the load on two generators.

Inspection reveals that excessive pressures exist in the HP drain main in that ship, partially as a result of unrepaired steam traps and orifices. Both HP drains and auxiliary exhaust are designed to go into the deaerating feed tank; however, excessive HP drain pressure results in all the auxiliary exhaust being returned to the engineroom (in that class ship).

The excessive amount of auxiliary exhaust precludes use of only the two auxiliary condensers associated with their turbogenerators, so the watch must admit auxiliary exhaust into the main condenser. This will cause the LP turbine to heat, so it becomes necessary to jack over the main engine. This requires lighting off a lube oil service pump to furnish oil pressure while the main engine is being turned by the jacking gear.

At the same time, admission of some of the auxiliary exhaust into the main condenser requires putting the main circulating pump, one of the main condensate pumps, and one set of the main air ejectors into operation. The boiler firing rate increases to support the larger steam demand.

Now the ship is burning more fuel. The number of engineroom watchstanders must increase. More equipment in use decreases the opportunity for maintenance of that equipment. Efficiency and economy suffer. Moreover, the situation will not improve until the engineers can attack the problem associated with excessive HP drain pressures on a systematic basis.

CHAPTER FOOTNOTE

In this chapter we have discussed the engineering inspection goals of safety, efficiency, and economy. "Ensign Benson," or any other shipmate with the interest, can become a competent engineering plant inspector. Propulsion plant reliability/engineering excellence is a "team effort," and an officer of the deck or prospective OOD can be a member of this team.

An officer of the deck who wants to become comfortable with engineering will find himself even more comfortable with an appreciation of basic inspection techniques when he finds himself assigned to a zone inspection of the engineering department.

CHAPTER REFERENCES

"Fireroom Heat-Stress Conditions and the Auxiliary Gland Leak-Off System", NavSea Journal, October 1974, pp. 17–18

Principles of Naval Engineering, NavPers 10788B

Index